Confronting the Challenges of Participatory Culture

II0658440

This report was made possible by grants from the John D. and Catherine T. MacArthur Foundation in connection with its grant making initiative on Digital Media and Learning. For more information on the initiative visit www.macfound.org.

Confronting the Challenges of Participatory Culture

Media Education for the 21st Century

Henry Jenkins (P. I.)

with Ravi Purushotma, Margaret Weigel, Katie Clinton, and Alice J. Robison

The MIT Press
Cambridge, Massachusetts
London, England

For information about special quantity discounts, please email special_sales@mitpress.mit.edu.

This book was set in Stone Serif and Stone Sans by the MIT Press.

Printed and bound in the United States of America.

Library of Congress Cataloging-in-Publication Data

Library of Congress Cataloging-in-Publication Data

Jenkins, Henry, 1958–
Confronting the challenges of participatory culture : media education for the 21st century / Henry Jenkins with Ravi Purushotma . . . [et al.].
    p. cm.—(The John D. and Catherine T. MacArthur Foundation reports on digital media and learning)
Includes bibliographical references.
ISBN 978-0-262-51362-3 (pbk. : alk. paper)
1. Media literacy. 2. Media literacy—Study and teaching. 3. Mass media and culture. 4. Mass media in education. I. Title.
P96.M4J46 2009   302.23—dc22   2009009102

10 9 8 7 6 5 4

Contents

Series Foreword

The John D. and Catherine T. MacArthur Foundation Reports on Digital Media and Learning, published by the MIT Press, present findings from current research on how young people learn, play, socialize, and participate in civic life. The Reports result from research projects funded by the MacArthur Foundation as part of its $50 million initiative in digital media and learning. They are published openly online (as well as in print) in order to support broad dissemination and to stimulate further research in the field.

Confronting the Challenges of Participatory Culture

Executive Summary

According to a recent study from the Pew Internet & American Life project,[1] more than one-half of all teens have created media content, and roughly one-third of teens who use the Internet have shared content they produced. In many cases, these teens are actively involved in what we are calling *participatory cultures*. A participatory culture is a culture with relatively low barriers to artistic expression and civic engagement, strong support for creating and sharing creations, and some type of informal mentorship whereby experienced participants pass along knowledge to novices. In a participatory culture, members also believe their contributions matter and feel some degree of social connection with one another (at the least, members care about others' opinions of what they have created).

Forms of participatory culture include:

Affiliations Memberships, formal and informal, in online communities centered around various forms of media, such as Friendster, Facebook, MySpace, message boards, metagaming, or game clans.

Expressions   Producing new creative forms, such as digital sampling, skinning and modding, fan videos, fan fiction, zines, or mash-ups.

Collaborative problem solving   Working together in teams, formal and informal, to complete tasks and develop new knowledge, such as through Wikipedia, alternative reality gaming, or spoiling.

Circulations   Shaping the flow of media, such as podcasting or blogging.

A growing body of scholarship suggests potential benefits from these forms of participatory culture, including opportunities for peer-to-peer learning, a changed attitude toward intellectual property, the diversification of cultural expression, the development of skills valued in the modern workplace, and a more empowered conception of citizenship. Access to this participatory culture functions as a new form of the hidden curriculum, shaping which youths will succeed and which will be left behind as they enter school and the workplace.

Some have argued that children and youths acquire these key skills and competencies on their own by interacting with popular culture. Three concerns, however, suggest the need for policy and pedagogical interventions:

The participation gap   The unequal access to the opportunities, experiences, skills, and knowledge that will prepare youths for full participation in the world of tomorrow.

The transparency problem   The challenges young people face in learning to recognize the ways that media shape perceptions of the world.

The ethics challenge The breakdown of traditional forms of professional training and socialization that might prepare young people for their increasingly public roles as media makers and community participants.

Educators must work together to ensure that all young Americans have access to the skills and experiences needed to become full participants, can articulate their understanding of how media shapes perceptions, and are socialized into the emerging ethical standards that should shape their practices as media makers and participants in online communities.

A central goal of this report is to shift the focus of the digital-divide discourse from questions of technological access to those of opportunities for participation and the development of cultural competencies and social skills needed for full involvement. Schools as institutions have been slow to react to the emergence of this new participatory culture; the greatest opportunity for change is currently found in after-school programs and informal learning communities. Schools and after-school programs must devote more attention to fostering what we call the *new media literacies:* a set of cultural competencies and social skills that young people need in the new media landscape. Participatory culture shifts the focus of literacy from individual expression to community involvement. The new literacies almost all involve social skills developed through collaboration and networking. These skills build on the foundation of traditional literacy and research, technical, and critical-analysis skills learned in the classroom.

The new skills include:

Play   The capacity to experiment with the surroundings as a form of problem solving.

Performance   The ability to adopt alternative identities for the purpose of improvisation and discovery.

Simulation   The ability to interpret and construct dynamic models of real-world processes.

Appropriation   The ability to meaningfully sample and remix media content.

Multitasking   The ability to scan the environment and shift focus onto salient details.

Distributed cognition   The ability to interact meaningfully with tools that expand mental capacities.

Collective intelligence   The ability to pool knowledge and compare notes with others toward a common goal.

Judgment   The ability to evaluate the reliability and credibility of different information sources.

Transmedia navigation   The ability to follow the flow of stories and information across multiple modalities.

Networking   The ability to search for, synthesize, and disseminate information.

Negotiation   The ability to travel across diverse communities, discerning and respecting multiple perspectives, and grasping and following alternative norms.

Fostering such social skills and cultural competencies requires a systemic approach to media education in the United States. Everyone involved in preparing young people to go out into the world has valuable contributions to help students acquire the skills necessary for becoming full participants in society. Schools, after-school programs, and parents have distinctive roles in encouraging and nurturing these skills.

Note

1. Amanda Lenhart and Mary Madden, *Teen Content Creators and Consumers* (Washington, DC: Pew Internet & American Life Project, 2005), http://www.pewinternet.org/PPF/r/166/report_display.asp.

The Needed Skills in the New Media Culture

If it were possible to define generally the mission of education, it could be said that its fundamental purpose is to ensure that all students benefit from learning in ways that allow them to participate fully in public, community, [creative,] and economic life.

—New London Group, "A Pedagogy of Multiliteracies: Designing Social Futures"[1]

Ashley Richardson was a middle-schooler when she ran for president of Alphaville. She wanted to control a government of more than 100 volunteer workers to make policies affecting thousands of people. She debated her opponent on National Public Radio and found herself in the center of a debate about the nature of citizenship, how to ensure honest elections, and the future of democracy in a digital age. Alphaville is the largest city in a popular multiplayer game, *The Sims Online*.[2]

Heather Lawver was 14 years old. She wanted to help other young people improve their reading and writing skills, so she established an online publication with a staff of more than 100 people across the world. As her project was embraced by teachers and integrated into their curriculum, she emerged as an important spokesperson in a national debate about intellectual property. The Web site Lawver created was a school newspaper for the fictional school Hogwarts, the setting for the popular Harry Potter books.[3]

Blake Ross was 14 years old when he was hired for a summer internship at Netscape. By that point, he had already developed computer programming skills and published his own Web site. Frustrated by many of the corporate decisions made at Netscape, Ross decided to design his own Web browser. Through the joint participation of thousands of other volunteer youths and adults working on his project worldwide, the Firefox Web browser was born. Today, Firefox enjoys more than 60 times as many users as Netscape Navigator. By age 19, Ross had the venture capital needed to launch his own start-up company. His interest in computing originally was sparked by playing the popular video game *Sim City*.[4]

Josh Meeter was about to graduate from high school when he completed the Claymation animation for *Awards Showdown*. He had negotiated with composer John Williams for the rights to use excerpts from his film scores. The film became widely circulated on the Web. By networking, Meeter was able to convince Stephen Spielberg to watch the film, and it was later featured on Spielberg's Dreamworks Web site. Meeter is now starting work on his first feature film.[5]

Richardson, Lawver, Ross, and Meeter are future politicians, activists, educators, writers, entrepreneurs, and media makers. The skills they acquired—learning how to campaign and govern; how to read, write, edit, and defend civil liberties; how to program computers and run a business; how to make a movie and find distribution—are the kinds of skills we might hope our best schools would teach. Yet none of these learning activities took place in schools. Indeed, many of these youths were frustrated with school; some dropped out and others chose to graduate early. They developed much of their skill and knowledge through their participation in the informal learning communities of fans and gamers.

Richardson, Lawver, Ross, and Meeter are exceptional individuals. In any given period, exceptional individuals will break rules and enjoy off-the-charts success—even at surprisingly young ages. But, Richardson, Lawver, Ross, and Meeter are not so exceptional.

According to a 2005 study conducted by the Pew Internet & American Life project,[6] more than half of all American teens— and 57 percent of teens who use the Internet—could be considered media creators. For the purpose of the study, a media creator is someone who has created a blog or Web page; posted original artwork, photography, stories, or videos online; or remixed online content into their own new creations. Most have done two or more of these activities. One-third of teens share what they create online with others, 22 percent have their own Web sites, 19 percent blog, and 19 percent remix online content.

Contrary to popular stereotypes, these activities are not restricted to white, suburban males. In fact, urban youths (40 percent) are somewhat more likely than their suburban (28 percent) or rural (38 percent) counterparts to be media creators. Girls aged 15–17 (27 percent) are more likely than boys their age (17 percent) to be involved with blogging or other social activities online. The Pew researchers found no significant differences in participation by race or ethnicity.

According to a 2005 study conducted by the Pew Internet & American Life project, more than half of all American teens—and 57 percent of teens who use the Internet—could be considered media creators.

If anything, the Pew study undercounts the number of American young people who are embracing the new participatory culture. The Pew study did not consider newer forms of expression, such as podcasting, game modding, or machinima. Nor did it count other forms of creative expression and appropriation, such as music sampling in the hip-hop community. These activities are highly technological, but they use tools and tap production and distribution networks neglected in the Pew study. The study also does not include even more widespread practices, such as computer or video gaming, that can require an extensive focus on constructing and performing as fictional personas. Our focus here is not on individual accomplishment but rather the emergence of a cultural context that supports widespread participation in the production and distribution of media.

Enabling Participation

While to adults the Internet primarily means the world wide web, for children it means email, chat, games—and here they are already content producers. Too often neglected, except as a source of risk, these communication and entertainment–focused activities, by contrast with the information-focused uses at the centre of public and policy agendas, are driving emerging media literacy. Through such uses, children are most engaged—multi-tasking, becoming proficient at navigation and manoeuvre so as to win, judging their participation and that of others, etc. . . . In terms of personal development, identity, expression and their social consequences—participation, social capital, civic culture—these are the activities that serve to network today's younger generation.

—Sonia Livingstone, *The Changing Natures and Uses of Media Literacy*[7]

Participatory Culture

For the moment, let's define participatory culture as one with

1. relatively low barriers to artistic expression and civic engagement,
2. strong support for creating and sharing creations with others,

3. some type of informal mentorship whereby what is known by the most experienced is passed along to novices,

4. members who believe that their contributions matter, and

5. members who feel some degree of social connection with one another (at the least, they care what other people think about what they have created).

Not every member must contribute, but all must believe they are free to contribute and that what they contribute will be appropriately valued.

In such a world, many will only dabble, some will dig deeper, and still others will master the skills that are most valued within the community. The community itself, however, provides strong incentives for creative expression and active participation. Historically, we have valued creative writing or art classes not only because they help to identify and train future writers and artists, but also because the creative process is valuable on its own; every child deserves the chance to express him- or herself through words, sounds, and images, even if most will never write, perform, or draw professionally. Having these experiences, we believe, changes the way youths think about themselves and alters the way they look at work created by others.

Most public-policy discussion of new media has centered on technologies—tools and their affordances. The computer is discussed as a magic black box with the potential to create a learn-

---

Participatory culture shifts the focus of literacy from individual expression to community involvement.

ing revolution (in the positive version) or a black hole that consumes resources that might be better devoted to traditional classroom activities (in the more critical version). Yet, as the epigraph above suggests, media operate in specific cultural and institutional contexts that determine how and why they are used. We may never know whether a tree makes a sound when it falls in a forest with no one around. But, clearly, a computer does nothing in the absence of a user. The computer does not operate in a vacuum. Injecting digital technologies into the classroom necessarily affects our relationship with every other communications technology, changing how we feel about what can or should be done with pencils and paper, chalk and black-board, books, films, and recordings.

Rather than dealing with each technology in isolation, we would do better to take an ecological approach, thinking about the interrelationship among different communication technologies, the cultural communities that grow up around them, and the activities they support. Media systems consist of communication technologies and the social, cultural, legal, political, and economic institutions, practices, and protocols that shape and surround them.[8] The same task can be performed with a range of different technologies, and the same technology can be

> Rather than dealing with each technology in isolation, we would do better to take an ecological approach, thinking about the interrelationship among different communication technologies, the cultural communities that grow up around them, and the activities they support.

deployed toward a variety of different ends. Some tasks may be easier with some technologies than with others, and thus the introduction of a new technology may inspire certain uses. Yet these activities become widespread only if the culture supports them, if they fill recurring needs at a particular historical juncture. The tools available to a culture matter, but what that culture chooses to do with those tools matters more.

The importance of culture's complex relationships with technologies is why we focus in this paper on the concept of participatory cultures rather than on interactive technologies. Inter-activity is a property of the technology,[9] while participation is a property of culture. Participatory culture is emerging as the culture absorbs and responds to the explosion of new media technologies that make it possible for average consumers to archive, annotate, appropriate, and recirculate media content in powerful new ways. A focus on expanding access to new technologies carries us only so far if we do not also foster the skills and cultural knowledge necessary to deploy those tools toward our own ends.

Participatory culture is emerging as the culture absorbs and responds to the explosion of new media technologies that make it possible for average consumers to archive, annotate, appropriate, and recirculate media content in powerful new ways. A focus on expanding access to new technologies carries us only so far if we do not also foster the skills and cultural knowledge necessary to deploy those tools toward our own ends.

We are using participation as a term that cuts across educational practices, creative processes, community life, and democratic citizenship. Our goals should be to encourage youths to develop the skills, knowledge, ethical frameworks, and self-confidence needed to be full participants in contemporary culture. Many young people are already part of this process through

Affiliations Memberships, formal and informal, in online communities centered around various forms of media, such as Friendster, Facebook, MySpace, message boards, metagaming, or game clans.

Expressions Producing new creative forms, such as digital sampling, skinning and modding, fan videos, fan fiction, zines, or mash-ups.

Collaborative problem-solving Working together in teams—formal and informal—to complete tasks and develop new knowledge, such as through *Wikipedia*, alternative reality gaming, or spoiling.

Circulations Shaping the flow of media, such as podcasting or blogging.

The MacArthur Foundation has launched an ambitious effort to document these activities and the roles they play in young people's lives. We do not want to preempt or duplicate that effort here. For the moment, it is sufficient to argue that each of these activities contains opportunities for learning, creative expression, civic engagement, political empowerment, and economic advancement.

Through these various forms of participatory culture, young people are acquiring skills that will serve them well in the future. Participatory culture is reworking the rules by which school, cultural expression, civic life, and work operate. A growing body of work has focused on the value of participatory culture and its long-term impact on children's understanding of themselves and the world around them.

Affinity Spaces

Many have argued that these new participatory cultures represent ideal learning environments. James Gee calls such informal learning cultures "affinity spaces,"[10] and explores why people learn more, participate more actively, and engage more deeply with popular culture than they do with the contents of their textbooks. Affinity spaces offer powerful opportunities for learning, Gee argues, because they are sustained by common endeavors that bridge differences—age, class, race, gender, and educational level—and because people can participate in various ways according to their skills and interests, because they depend on peer-to-peer teaching with each participant constantly motivated to acquire new knowledge or refine their existing skills, and because they allow each participant to feel like an expert while tapping the expertise of others. For example, Rebecca Black finds that the "beta-reading" (or editorial feedback) provided by online fan communities helps contributors grow as writers, not only helping them master the basic building blocks of sentence construction and narrative structure, but also pushing them to be close readers of the works that inspire them.[11] Participants in the beta-reading process learn both by receiving feedback on their own work and by giving feedback to others, creating an ideal peer-to-peer learning community.

Affinity spaces are distinct from formal educational systems in several ways. While formal education is often conservative, the informal learning within popular culture is often experimental. While the formal is static, the informal is innovative. The structures that sustain informal learning are more provisional; those supporting formal education are more institutional. Informal learning communities can evolve to respond to short-term needs and temporary interests, whereas the institutions supporting public education have changed little despite decades of school reform. Informal learning communities are ad hoc and localized; formal educational communities are bureaucratic and increasingly national in scope. We can move in and out of informal learning communities if they fail to meet our needs; we enjoy no such mobility in our relations to formal education.

Affinity spaces are also highly generative environments from which new aesthetic experiments and innovations emerge. A 2005 report on *The Future of Independent Media* argued that this kind of grassroots creativity was an important engine of cultural transformation:

The media landscape will be reshaped by the bottom-up energy of media created by amateurs and hobbyists as a matter of course. This bottom-up energy will generate enormous creativity, but it will also tear apart some of the categories that organize the lives and work of media makers. . . . A new generation of media-makers and viewers are emerging which could lead to a sea change in how media is made and consumed.[12]

This report celebrates a world in which everyone has access to the means of creative expression and the networks supporting artistic distribution. The Pew study suggests something more:

young people who create and circulate their own media are more likely to respect the intellectual property rights of others because they feel a greater stake in the cultural economy.[13] Both reports suggest we are moving away from a world in which some produce and many consume media toward one in which everyone has a more active stake in the culture that is produced.

David Buckingham argues that young people's lack of interest in news and their disconnection from politics reflects their perception of disempowerment. "By and large, young people are not defined by society as political subjects, let alone as political agents. Even in the areas of social life that affect and concern them to a much greater extent than adults—most notably education—political debate is conducted almost entirely 'over their heads.'"[14] Politics, as constructed by the news, becomes a spectator sport, something we watch but not do. Yet the new participatory culture offers many opportunities for youths to engage in civic debates, participate in community life, and even become political leaders, even if sometimes only through the "second lives" offered by massively multiplayer games or online fan communities.

Empowerment comes from making meaningful decisions within a real civic context: we learn the skills of citizenship by becoming political actors and gradually coming to understand

> We are moving away from a world in which some produce and many consume media toward one in which everyone has a more active stake in the culture that is produced.

the choices we make in political terms. Today's children learn through play the skills they will apply to more serious tasks later. The challenge is how to connect decisions made in the context of our everyday lives with the decisions made at local, state, or national levels. The step from watching television news to acting politically seems greater than the transition from being a political actor in a game world to acting politically in the "real world."

Participating in these affinity spaces also has economic implications. We suspect that young people who spend more time playing within these new media environments will feel greater comfort interacting with one another via electronic channels, will have greater fluidity in navigating information landscapes, will be better able to multitask and make rapid decisions about the quality of information they are receiving, and will be able to collaborate better with people from diverse cultural backgrounds. These claims are borne out by research conducted by Beck and Wade into the ways that early game-play experiences affect subsequent work habits and professional activities.[15] Beck and Wade conclude that gamers were more open to taking risks and engaging in competition but also more open to collaborating with others and more willing to revise earlier assumptions.

This focus on the value of participating within the new media culture stands in striking contrast to recent reports from the Kaiser Family Foundation that have bemoaned the amount of time young people spend on "screen media."[16] The Kaiser reports collapse a range of different media-consumption and media-production activities into the general category of "screen time" without reflecting very deeply on the different degrees of social connectivity, creativity, and learning involved. We do not mean to dismiss the very real concerns they raise: that mediated experience may squeeze out time for other learning activities; that contemporary children often lack access to real-world play spaces, with adverse health consequences; that adults may inadequately supervise and interact with children regarding the media they consume (and produce); and that the moral values and commercialization in much contemporary entertainment may be harmful. Yet the focus on negative effects of media consumption offers an incomplete picture. These accounts do not appropriately value the skills and knowledge young people are gaining through their involvement with new media, and, as a consequence, they may mislead us about the roles teachers and parents should play in helping children learn and grow.

Why We Should Teach Media Literacy: Three Core Problems

Some defenders of the new digital cultures have acted as though youths can simply acquire these skills on their own without adult intervention or supervision. Children and youths do indeed know more about these new media environments than most parents and teachers. In fact, we do not need to protect them so much as engage them in critical dialogues that help them to articulate more fully their intuitive understandings of these experiences. To say that children are not victims of media is not to say that they, any more than anyone else, have fully mastered the complex and still-emerging social practices.

There are three core flaws with the laissez-faire approach. The first is that it does not address the fundamental inequalities in young people's access to new media technologies and the opportunities for participation they represent (what we call the *participation gap*). The second is that it assumes that children are actively reflecting on their media experiences and thus can articulate what they learn from their participation (the *transparency problem*). The third is that it assumes children, on their own, can develop the ethical norms needed to cope with a complex and diverse social environment online (the *ethics challenge*).

Any attempt to provide meaningful media education in the age of participatory culture must begin by addressing these three core concerns.

The Participation Gap

Cities around the country are providing wireless Internet access for their residents. Some cities, such as Tempe, Arizona, charge users a fee; others, such as Philadelphia, Boston, and Cambridge, plan to provide high-speed wireless Internet access free of charge. In an interview on PBS's *Nightly News Hour* in November 2005, Philadelphia mayor John Street spoke of the link between Internet access and educational achievement:

Philadelphia will allow low-income families, families that are on the cusp of their financial capacity, to be able to be fully and completely connected. We believe that our public school children should be—their families have to be connected or else they will fall behind, and, in many cases, never catch up.[17]

Philadelphia's Emergency People's Shelter (EPS) is ahead of the curve; the nonprofit group's free network access serves shelter residents and the surrounding neighborhood. Gloria Guard of EPS said,

What we realized is if we can't get computers into the homes of our constituents and our neighbors and of this neighborhood, there are children in those households who will not be able to keep up in the marketplace. They won't be able to keep up with their schoolmates. They won't be able to even apply for college. We thought it was really important to get computer skills and connection to the Internet into as many homes as possible.[18]

However, simply passing out technology is not enough. Expanding access to computers will help bridge some of the gaps between digital haves and have nots, but only in a context in which free Wi-Fi is coupled with new educational initiatives to help youths and adults learn how to use those tools effectively.

Throughout the 1990s, the country focused enormous energy on combating the digital divide in technological access. The efforts have ensured that most American youths have at least minimal access to networked computers at school or in public libraries. However, as a 2005 report on children's online experience in the United Kingdom concluded,

> No longer are children and young people only or even mainly divided by those with or without access, though "access" is a moving target in terms of speed, location, quality and support, and inequalities in access do persist. Increasingly, children and young people are divided into those for whom the Internet is an increasingly rich, diverse, engaging and stimulating resource of growing importance in their lives and those for whom it remains a narrow, unengaging, if occasionally useful, resource of rather less significance.[19]

What a person could accomplish with an outdated machine in a public library with mandatory filtering software and no opportunity for storage or transmission pales in comparison to what the same person could accomplish with a home computer with unfettered Internet access, high bandwidth, and continuous connectivity.[20] Our school systems' inability to close this participation gap has negative consequences for everyone involved. On the one hand, those youths who are most advanced in media literacies are often stripped of their technologies and robbed of their best techniques for learning in an effort to ensure a uniform experience for all in the classroom. On the other hand, many youths who have had no exposure to these new participatory cultures outside school find themselves struggling to keep up with their peers.

Wartella, O'Keefe, and Scantlin reached a similar conclusion:

Closing the digital divide will depend less on technology and more on providing the skills and content that is most beneficial. . . . Children who have access to home computers demonstrate more positive attitudes towards computers, show more enthusiasm and report more enthusiasm and ease when using computers than those who do not.[21]

More often than not, those youths who have developed the most comfort with the online world are the ones who dominate classroom use of computers, pushing aside less technically skilled classmates. We would be wrong, however, to see this as a simple binary of youths who have technological access and those who do not. Wartella and her coauthors note, for example, that game systems make their way into a growing number of working-class homes, even if laptops and personal computers do not. Working-class youths may have access to some of the benefits of play described here, but they may still lack the ability to produce and distribute their own media.

In a 2005 report prepared for the MacArthur Foundation, Lyman finds that children's experiences online are shaped by a range of social factors, including class, age, gender, race, nationality, and point of access. He notes, for example, that middle-class youths are more likely to rely on resources and assistance from peers and family within their own homes and thus seem more autonomous at school than working-class children, who often must rely more heavily on teachers and peers to make up for a lack of experience at home. The middle-class children thus seem "naturally" superior in their use of technology, further amplifying their own self-confidence in their knowledge.[22]

Historically, those youths who had access to books or classical recordings in their homes, whose parents took them to concerts or museums, or who engaged in dinner conversation developed—almost without conscious consideration—skills that helped them perform well in school. Those experiences, which were widespread among the middle class and rare among the working class, became a kind of class distinction that shaped how teachers perceived students. These new forms of cultural participation may be playing a similar role. These activities shape the skills and knowledge students bring into the classroom and, in this fashion, determine how teachers and peers perceive these students. Castells tells us about youths who are excluded from these experiences: "Increasingly, as computer use is ever less a lifestyle option, ever more an everyday necessity, inability to use computers or find information on the web is a matter of stigma, of social exclusion, revealing not only changing social norms but also the growing centrality of computers to work, education and politics."[23]

Writing on how contemporary industry values our "portfolios" as much as our knowledge, Gee suggests that what gives elite teens their head start is their capacity to

pick up a variety of experiences (e.g., the "right" sort of summer camps, travel, and special activities), skills (not just school-based skills, but a wide variety of interactional, aesthetic, and technological skills), and achievements (honors, awards, projects) in terms of which they can help to define themselves as worthy of admission to elite educational institutions and worthy of professional success later in life.[24]

They become adept at identifying opportunities for leadership and accomplishment; they adjust quickly to new situations, embrace new roles and goals, and interact with people of diverse backgrounds. Even if these opportunities were not formally valued by our educational institutions or listed on a resume when applying for a job, the skills and self-confidence gathered by moving across all of these online communities surely would manifest themselves in other ways, offering yet another leg up to youths on one side and another disadvantage to youths on the opposite side of the participation gap.

The Transparency Problem

Although youths are becoming more adept at using media as resources (for creative expression, research, social life, etc.), they often are limited in their ability to examine the media themselves. Sherry Turkle was among the first to call attention to this transparency problem:

Games such as *SimLife* teach players to think in an active way about complex phenomena (some of them "real life," some of them not) as dynamic, evolving systems. But they also encourage people to get used to manipulating a system whose core assumptions they do not see and which may or may not be "true."[25]

Not everyone agrees. In an essay on the game *SimCity*, Ted Friedman contends that game players seek to identify and exploit the rules of the system in order to beat the game.[26] The antagonistic relationship between player and game designer means that game players may be more suspicious of the rules structuring their experiences than are the consumers of many other kinds of media. Conversations about games expose flaws in games' construction, which also may lead to questions about the games' governing assumptions. Subsequent games have, in fact, allowed players to reprogram the core models. There is a difference between trying to master the rules of the game and recognizing the ways those rules structure our perception of reality, though. It may be much easier to see what is in the game than to recognize what is missing.

This issue of transparency crops up regularly in the first wave of field reports on the pedagogical use of games. Karen Schrier developed a location-specific game for teaching American history, which was played in Lexington, Massachusetts; her game was designed to encourage reflection on competing and contradictory accounts of who fired the first shot of the American Revolution.[27] The project asked students to experience the ways historians interpret evidence and evaluate competing truths. Such debates emerged spontaneously around the game-play experience. Yet Schrier was surprised by another phenomenon: the young people took the game's representation of historical evidence at face value, acting as if all of the information in the game were authentic.

Schrier offers several possible explanations for this transparency problem, ranging from the legacy of textbook publishing, where instructional materials did not encourage users to question their structuring or their interpretation of the data, to the tendency to "suspend our disbelief" in order to have a more immersive play experience. Kurt Squire found similar patterns when he sought to integrate the commercial game *Civilization III* into world-history classes. Students were adept at formulating "what if" hypotheses, which they tested through their game play, yet they lacked a vocabulary to critique how the game itself constructed history, and they had difficulty imagining how other games might represent the same historical processes in different terms. In both cases, students were learning how to read information from and through games, but they were not yet learning how to read games as texts that were constructed with their own aesthetic norms, genre conventions, ideological biases, and codes of representation.[28] These findings suggest the importance of coupling the pedagogical use of new media technologies with a greater focus on media literacy.

These concerns about the transparency of games, even when used in instructional contexts, are closely related to concerns about how young people (or indeed, any of us) assess the quality of information we receive. As Renee Hobbs has suggested, "Determining the truth value of information has become increasingly difficult in an age of increasing diversity and ease of access to information."[29] More recent work by the Harvard Good Works Project has found that issues of format and design are often more important than issues of content in determining

---

Students were adept at formulating "what if" hypotheses, which they tested through their game play, yet they lacked a vocabulary to critique how the game itself constructed history, and they had difficulty imagining how other games might represent the same historical processes in different terms.

how much credibility young people attach to the content of a particular Web site.[30] This research suggests some tendency to read "professional" sites as more credible than "amateur" materials, although students lack a well-developed set of standards for distinguishing between the two. In her recent book *The Internet Playground*, Ellen Seiter expresses concern that young people are finding it increasingly difficult to separate commercial from noncommercial content in online environments: "The Internet is more like a mall than a library; it resembles a gigantic public relations collection more than it does an archive of scholars."[31]

Increasingly, content comes to us already branded, already shaped through an economics of sponsorship, if not overt advertising. We do not know how much these commercial interests influence what we see and what we do not see. Commercial interests even shape the order of listings on search engines in ways that are often invisible to those who use them. Increasingly, opportunities to participate online are branded such that even when young people produce and share their own media, they do so under terms set by commercial interests. Children, Seiter found, often had trouble identifying advertising practices in the popular *Neopets* site, in part because the product references were so integrated into the game. The children were used to a world where commercials stood apart from the entertainment content and thus they equated branding with banner advertisements. This is where the transparency issue becomes especially dangerous. Seiter concludes, "The World Wide Web is a more aggressive and stealthy marketeer to children than television ever was, and children need as much

Children often had trouble identifying advertising practices in the popular *Neopets* site, in part because the product references were so integrated into the game. The children were used to a world where commercials stood apart from the entertainment content and thus they equated branding with banner advertisements.

information about its business practices as teachers and parents can give them."[32] Children need a safe space within which they can master the skills they need as citizens and consumers as they learn to parse messages from self-interested parties, and where they can separate fact from falsehood as they begin to experiment with new forms of creative expression and community participation.

The Ethics Challenge

In *Making Good: How Young People Cope with Moral Dilemmas at Work*, Wendy Fischman and her coauthors discuss how young journalists learn the ethical norms that will define their future professional practice. These writers, they find, acquired their skills most often by writing for high school newspapers. For the most part, the authors suggest, student journalists worked in highly cohesive and insulated settings. Their work was supervised, for better or worse, by a range of adult authorities, some interested in promoting the qualities of good journalism, some concerned with protecting the reputation of the school. Their work was free of commercial constraints and sheltered from outside exposure. The ethical norms and professional practices they were acquiring were well understood by the adults around them.[33]

Now, consider how few of those qualities might be applied to the emerging participatory cultures. In a world in which the line between consumers and producers is blurring, young people are finding themselves in situations that no one would have anticipated a decade or two ago. Their writing is much more open to the public and can have more far-reaching consequences. Young people are creating new modes of expression that are poorly understood by adults, and as a result they receive little to no guidance or supervision. The ethical implications of these emerging practices are fuzzy and ill-defined. Young people are discovering that information they put online to share with their friends can bring unwelcome attention from strangers.

In professional contexts, professional organizations are the watchdog of ethical norms. Yet in more casual settings, there is seldom a watchdog at all. No established set of ethical guidelines shapes the actions of bloggers and podcasters, for example. How should teens decide what they should or should not post about themselves or their friends on *LiveJournal* or *MySpace*? Different online communities have their own norms about what information should remain within the group and what can be circulated more broadly, and many sites depend on self-disclosure to police whether the participants are children or adults. Yet many young people seem willing to lie to access those communities.

Ethics become much murkier in game spaces, where invented identities are assumed and actions are fictive, designed to allow broader rein to explore (sometimes darker) fantasies. That said, unwritten and often imperfectly shared norms exist about acceptable or unacceptable conduct. Essays such as Julian Dibbel's "A Rape in Cyberspace," Henry Jenkins's "Playing Politics in Alphaville," and always_black's "Bow Nigger" offer reminders that participants in these worlds understand the same experiences in very different terms and follow different ethical norms as they face off against each other.[34]

In *Making Good*, Fischman and coauthors found that high school journalists felt constrained by the strong social ties in their high school and were unwilling to publish some articles they believed would be received negatively by their peers or that might disrupt the social dynamics of their society. What constraints, if any, apply in online realms? Do young people feel that same level of investment in their gaming guilds or their fan communities? Or do the abilities to mask real identities or move from one community to another mean there are fewer immediate consequences for antisocial behavior?

One important goal of media education should be to encourage young people to become more reflective about the ethical choices they make as participants and communicators and about the impact they have on others. In the short run, we may have to accept that cyberspace's ethical norms are in flux: we are taking part in a prolonged experiment in what happens when barriers of entry into a communication landscape become lower. For the present moment, asking and working through questions of ethical practices may be more valuable than the answers produced because the process will help everyone to recognize and articulate the different assumptions that guide their behavior.

As we think about meaningful pedagogical intervention, we must keep in mind three core concerns:

• How do we ensure that every child has access to the skills and experiences needed to become a full participant in the social, cultural, economic, and political future of our society?
• How do we ensure that every child has the ability to articulate his or her understanding of how media shapes perceptions of the world?
• How do we ensure that every child has been socialized into the emerging ethical standards that should shape their practices as media makers and as participants in online communities?

To address these challenges, we must rethink which core skills and competencies we want our children to acquire in their learning experiences. The new participatory culture places new emphasis on familiar skills that have long been central to American education; it also requires teachers to pay greater attention to the social skills and cultural competencies that are emerging in the new media landscape. In the next sections, we provide a framework for thinking about the type of learning that should occur if we are to address the participation gap, the transparency problem, and the ethics challenges.

What Should We Teach? Rethinking Literacy

Adolescents need to learn how to integrate knowledge from multiple sources, including music, video, online databases, and other media. They need to think critically about information that can be found nearly instantaneously throughout the world. They need to participate in the kinds of collaboration that new communication and information technologies enable, but increasingly demand. Considerations of globalization lead us toward the importance of understanding the perspective of others, developing a historical grounding, and seeing the interconnectedness of economic and ecological systems.

—Bertram C. Bruce, "Diversity and Critical Social Engagement: How Changing Technologies Enable New Modes of Literacy in Changing Circumstances"[35]

A definition of *twenty-first century literacy* offered by the New Media Consortium is "the set of abilities and skills where aural, visual, and digital literacy overlap. These include the ability to understand the power of images and sounds, to recognize and use that power, to manipulate and transform digital media, to distribute them pervasively, and to easily adapt them to new forms."[36] We would modify this definition in two ways. First, textual literacy remains a central skill in the twenty-first century. Youths must expand their required competencies, not push aside old skills to make room for the new. Second, new media literacies should be considered a social skill.

> The new literacies almost all involve social skills developed
> through collaboration and networking. These skills build on the
> foundation of traditional literacy, research skills, technical skills,
> and critical-analysis skills taught in the classroom.

New media literacies include the traditional literacy that
evolved with print culture as well as the newer forms of literacy
within mass media and digital media. Much writing about
twenty-first century literacies seems to assume that communicating through visual, digital, or audiovisual media will displace
reading and writing. We fundamentally disagree. Before students can engage with the new participatory culture, they must
be able to read and write. Just as the emergence of written language changed oral traditions and the emergence of printed
texts changed our relationship to written language, the emergence of new digital modes of expression changes our relationship to printed texts. In some ways, as researchers such as Black
and Jenkins have argued, the new digital cultures provide support systems to help youths improve their core competencies as
readers and writers.[37] They may provide opportunities, for
example through blogs or live journals, for young people to
receive feedback on their writing and to gain experience in
communicating with a larger public, experiences that might
once have been restricted to student journalists. Even traditional literacies must change to reflect the media change taking
place. Youths must expand their required competencies, not
push aside old skills to make room for the new.

Before students can engage with the new participatory culture, they must be able to read and write. Youth must expand their required competencies, not push aside old skills to make room for the new.

Beyond core literacy, students need research skills. Among other things, they need to know how to access books and articles through a library; to take notes on and integrate secondary sources; to assess the reliability of data; to read maps and charts; to make sense of scientific visualizations; to grasp what kinds of information are being conveyed by various systems of representation; to distinguish between fact and fiction, fact and opinion; and to construct arguments and marshal evidence. If anything, these traditional skills assume even greater importance as students venture beyond collections that have been screened by librarians and into the more open space of the Web. Some of these skills have traditionally been taught by librarians who, in the modern era, are reconceptualizing their role less as curators of bounded collections and more as information facilitators who can help users find what they need, online or offline, and can cultivate good strategies for searching material.

Students also need to develop technical skills. They need to know how to log on, to search, to use various programs, to focus a camera, to edit footage, to do some basic programming, and so forth. Yet, to reduce the new media literacies to technical skills would be a mistake on the order of confusing penmanship with composition. Because the technologies are undergoing

To reduce the new media literacies to technical skills would be a mistake on the order of confusing penmanship with composition. Schools have so far conceived of the challenges of digital media primarily in these technical terms, with the computer lab displacing the typing classroom. Too often, however, this training occurs in a vacuum, cut off from larger notions of literacy or research.

such rapid change, it is probably impossible to codify which technologies or techniques students must know.

As media literacy advocates have claimed during the past several decades, students also must acquire a basic understanding of the ways media representations structure our perceptions of the world, the economic and cultural contexts within which mass media is produced and circulated, the motives and goals that shape the media they consume, and alternative practices that operate outside the commercial mainstream. Such groups have long called for schools to foster a critical understanding of media as one of the most powerful social, economic, political, and cultural institutions of our era. What we are calling here the *new media literacies* should be taken as an expansion of, rather than a substitution for, the mass media literacies.

Which New Skills Matter? New Social Skills and Cultural Competencies

All of these skills are necessary—even essential—but they are not sufficient, which brings us to our second point about the

> The new media literacies should be seen as social skills, as ways of interacting within a larger community, and not simply as individualized skills to be used for personal expression.

notion of twenty-first century literacy: the new media literacies should be seen as social skills, as ways of interacting within a larger community, and not simply as individualized skills to be used for personal expression. The social dimensions of literacy are acknowledged in the New Media Consortium's report only in terms of the distribution of media content.[38] We must push further by talking about how meaning emerges collectively and collaboratively in the new media environment and how creativity operates differently in an open-source culture based on sampling, appropriation, transformation, and repurposing.

The social production of meaning is more than individual interpretation multiplied; it represents a qualitative difference in the ways we make sense of cultural experience, and in that sense it represents a profound change in how we understand literacy. In such a world, youths need skills for working within social networks, for pooling knowledge within a collective intelligence, for negotiating across cultural differences that shape the governing assumptions in different communities, and for reconciling conflicting bits of data to form a coherent picture of the world around them.

We must integrate these new knowledge cultures into our schools, not only through group work but also through long-distance collaborations across different learning communities.

Students should discover what it is like to contribute their own expertise to a process that involves many intelligences, a process they encounter readily in their participation in fan discussion lists or blogging. Indeed, this disparate collaboration may be the most radical element of new literacies: they enable collaboration and knowledge sharing with large-scale communities that may never interact in person. Schools currently are still training autonomous problem solvers, while as students enter the workplace they are increasingly being asked to work in teams, drawing on different sets of expertise, and collaborating to solve problems.

Changes in the media environment are altering our understanding of literacy and requiring new habits of mind, new ways of processing culture and interacting with the world around us. We are just beginning to identify and assess these emerging sets of social skills and cultural competencies. We have only a broad sense of which competencies are most likely to matter as young people move from the realms of play and education and into the adult world of work and society. What follows, then, is a provisional list of eleven core skills needed to participate within the new media landscape. These skills have been identified both by reviewing the existing body of scholar-

---

Youths need skills for working within social networks, for pooling knowledge within a collective intelligence, for negotiating across cultural differences that shape the governing assumptions in different communities, and for reconciling conflicting bits of data to form a coherent picture of the world around them.

Schools currently are still training autonomous problem solvers, while as students enter the workplace they are increasingly being asked to work in teams, drawing on different sets of expertise, and collaborating to solve problems.

ship on new media literacies and by surveying the forms of informal learning taking place in the participatory culture. As suggested above, mastering these skills remains a key step in preparing young people "to participate fully in public, community, [creative,] and economic life."[39] In short, these are skills some youths are learning through participatory culture, but they are also skills that all youths need to learn if they are going to be equal participants in the world of tomorrow. We identify a range of activities that might be deployed in schools or after-school programs, across a range of disciplines and subject matter, to foster these social skills and cultural competencies. These activities are by no means an exhaustive list but rather are simply illustrations of the kind of work already being done in each area. One goal of this report is to challenge those who have responsibility for teaching our young people to think more systematically and creatively about the many different ways they might build these skills into their day-to-day activities in ways that are appropriate to the content they are teaching.

Core Media Literacy Skills

Play: The Capacity to Experiment with the Surroundings as a Form of Problem Solving

Play, as psychologists and anthropologists have long recognized, is key in shaping children's relationships to their bodies, tools, communities, surroundings, and knowledge. Most of children's earliest learning comes through playing with the materials at hand. Through play, children try on roles, experiment with culturally central processes, manipulate core resources, and explore their immediate environments. As they grow older, play can motivate other forms of learning.

Mary Louise Pratt describes what her son and his friend learned through baseball card collecting:

Sam and Willie learned a lot about phonics that year by trying to decipher surnames on baseball cards, and a lot about cities, states, heights, weights, places of birth, stages of life. . . . And baseball cards opened the door to baseball books, shelves and shelves of encyclopedias, magazines, histories, biographies, novels, books of jokes, anecdotes, cartoons, even poems. . . . Literacy began for Sam with the newly pronounceable names on the picture cards and brought him what has been easily the broadest, most varied, most rewarding, and most integrated experience of his 13-year life.[40]

Pratt's account suggests this playful activity motivated three very different kinds of learning. First, the activity itself demanded certain skills and practices that had clear payoffs for academic subjects. For example, working out batting averages gave Sam an occasion to rehearse his math skills, arranging his cards introduced him to the process of classification, and discussing the cards gave him reason to work on his communication skills. On another level, the cards provided a scaffold that motivated and shaped his acquisition of other forms of school knowledge. The cards inspired Sam to think about the cities where the teams were located and acquire map-reading skills. The history of baseball provided a context through which to understand twentieth-century American history. The interest in stadiums introduced some basics of architecture. Third, Sam developed a sense of himself as a learner: "He learned the meaning of expertise, of knowing about something well enough that you can start a conversation with a stranger and feel sure of holding your own."[41]

Game designer Scott Osterweill (*The Logical Journey of the Zoobinis*) has described the mental attitude that surrounds play as highly conducive for learning:

When children are deep at play they engage with the fierce, intense attention that we'd like to see them apply to their schoolwork. Interestingly enough, no matter how intent and focused a child is at that play, maybe even grimly determined they may be at that game play, if you asked them afterwards, they will say that they were having fun. So, the fun of game play is not non-stop mirth but rather the fun of engaging of attention that demands a lot of you and rewards that effort. I think most good teachers believe that in the best moments, classroom learning can be the same kind of fun. But a game is a moment when the kid gets to have that in spades, when the kid gets to be focused and intent and hardworking and having fun at the same time.[42]

You will note here a shift in emphasis from fun (which in our sometimes still-puritanical culture gets defined as the opposite of seriousness) to engagement. When individuals play games, a fair amount of what they end up doing is not especially fun at the moment. It can be a grind, not unlike homework. The effort allows the person to master skills, collect materials, or put things in their proper place in anticipation of a payoff down the line. The key is that this activity is deeply motivated. The individual is willing to go through the grind because there is a goal or purpose that matters to the person. When that happens, individuals are engaged, whether that be the engagement in professional lives or the learning process or the engagement that some find through playing games. For the current generation, games may represent the best way of tapping that sense of engagement with learning.

To date, much of the discussion of games and education has considered games as a tool to motivate youths to learn other kinds of content (Pratt's move from baseball cards to geogra-

---

For the current generation, games may represent the best way of tapping a sense of engagement with learning.

phy), but there is a growing recognition that play itself—as a means of exploring and processing knowledge and of problem solving—may be a valuable skill children should master in preparation for subsequent roles and responsibilities in the adult world.

Part of what makes play valuable as a mode of problem solving and learning is that it lowers the emotional stakes of failing: players are encouraged to suspend some of the real-world consequences of the represented actions, to take risks and learn through trial and error. The underlying logic is one of die and do over. As Gee has noted, children often feel locked out of the worlds described in their textbooks through the depersonalized and abstract prose used to describe them.[43] Games construct compelling worlds for players to move through. Players feel a part of those worlds and have some stake in the events unfolding. Games do not only provide a rationale for learning: what players learn is put to use immediately to solve compelling problems with real consequences in the world of the game. Game designer Will Wright (*SimCity*, *The Sims*) has argued:

In some sense, a game is nothing but a set of problems. We're actually selling people problems for 40 bucks a pop. . . . And the more interesting games in my opinion are the ones that have a larger solution space. In other words, there's not one specific way to solve a puzzle, but, in fact, there's an infinite range of solutions. . . . The game world becomes an external artifact of their internal representation of the problem space.[44]

For Wright, the players' hunger for challenge and complexity motivates them to pick up the game in the first place.

Games follow something akin to the scientific process. Players are asked to make their own discoveries and then apply what

Children often feel locked out of the worlds described in their textbooks through the depersonalized and abstract prose used to describe them. Games construct compelling worlds for players to move through. Players feel a part of those worlds and have some stake in the events unfolding.

they learn to new contexts. No sooner does a player enter a game than he or she begins by identifying core conditions and looking for problems that must be addressed. On the basis of the available information, the player poses a certain hypothesis about how the world works and the best ways of bringing its properties under his or her control. The player tests and refines that hypothesis through actions in the game, which either fail or succeed. The player refines the model of the world as he or she goes. More sophisticated games allow the player to do something more, to experiment with the properties of the world, to frame new possibilities, which involves manipulating relevant variables and seeing what happens. Meta-gaming, the discourse that surrounds games, provides a context for players to reflect on and articulate what they have learned through the game. Here, for example, is how Kurt Squire describes the meta-gaming that occurs with *Civilization III:*

Players enroll as advanced players, having spent dozens, if not hundreds, of hours with the game and having mastered its basic rules. As players begin to identify and exploit loopholes, they propose and implement changes to the games' rules, identify superior strategies, and invent new game rule systems, including custom modifications and scenarios.[45]

Some have expressed skepticism that schools should or could teach young people how to play. This resistance reflects the confusion between play as a source of fun and play as a form of engagement. Play in the context argued here is a mode of active engagement, one that encourages experimentation and risk taking, one that views the process of solving a problem as important as finding the answer, one that offers clearly defined goals and roles that encourage strong identifications and emotional investments. This form of play is closely related to two other important skills: simulation and performance.

What Might Be Done

Educators (in school and out) have been tapping into play as a skill by encouraging free-form experimentation and open-ended speculation.

• History teachers ask students to entertain alternative history scenarios, speculating on what might have happened if Germany had won World War II or if Native Americans had colonized Europe. Such questions can lead to productive explorations centering on why and how certain events occurred, and what effects they had. Such questions also have no right and wrong answers: they emphasize creative thinking rather than memorization, they allow diverse levels of engagement, they allow students to feel less intimidated by adult expertise, and they also lend themselves to the construction of arguments and the mobilization of evidence.

• Art and design students are turned loose with a diverse array of everyday materials and encouraged to use them to solve a specified design problem. Such activities encourage students to revisit familiar materials and everyday objects with fresh perspectives, to think through common problems from multiple directions, and to respect alternative responses to the same challenge. This approach is closely associated with the innovative design work of Ideo, a Palo Alto consultancy, but can also be seen in various reality television programs, such as *Project Runway* or *Iron Chef*, that require contestants to adopt distinctive and multiple approaches to shared problems.

• Games offer the potential for learning through a new form of direct experience. Physics teachers use the game *Supercharged*, which was developed as part of the MIT Games to Teach initiative, to help students better understand core principles of electromagnetism. Using the game as a means for learning the laws of electromagnetism through first-hand experience, students navigate electromagnetic mazes by planting electrical charges that attract or repel their vehicles. Teachers can then build on this intuitive and experiential learning in the classroom, introducing equations, diagrams, or visualizations that help the students better understand the underlying principles that they are deploying before sending them back to play through the levels again and improve their performance.

Simulation: The Ability to Interpret and Construct Dynamic Models of Real-World Processes

New media provides powerful new ways of representing and manipulating information. New forms of simulation expand our cognitive capacities, allowing us to deal with larger bodies of information, to experiment with more complex configurations of data, and to form hypotheses quickly and test them against different variables in real time. The emergence of systems-based thinking has arisen hand in hand with the development of digital simulations. Across a range of academic and professional fields, simulations can be effective in representing acquired knowledge or in testing emerging theories. Because simulations are dynamic, and because they are governed by the systematic application of grounding assumptions, they can be a tool for discovery as researchers observe the emergent properties of these virtual worlds. We learn through simulations by a process of trial and error: new discoveries lead researchers to refine their models by tweaking particular variables and trying out different contingencies. Educators have always known that students learn more through direct observation and experimentation than from reading about something in a textbook or listening to a lecture. Simulations broaden the kinds of experiences users can have with compelling data, giving them a chance to see and do things that would be impossible in the real world.

Contemporary video games allow youths to play with sophisticated simulations and, in the process, to develop an intuitive understanding of how we might use simulations to test our assumptions about the way the world works. John Seely Brown,

> Educators have always known that students learn more through direct observation and experimentation than from reading about something in a textbook or listening to a lecture. Simulations broaden the kinds of experiences users can have.

former head of Xerox Parc, tells the story of sixteen-year-old Colin, whose understanding of the ancient world had been shaped by the game *Caesar III*:

Colin said: "I don't want to study Rome in high school. Hell, I build Rome every day in my on-line game." . . . Of course, we could dismiss this narrative construction as not really being a meaningful learning experience, but a bit later he and his dad were engaged in a discussion about the meaningfulness of class distinctions—lower, middle, etc.— and his dad stopped and asked him what class actually means to him. Colin responded:
"Well, it's how close you are to the Senate."
"Where did you learn that, Colin?"
"The closer you are physically to the Senate building, the plazas, the gardens, or the Triumphal Arch raises the desirability of the land, makes you upper class and produces plebeians. It's based on simple rules of location to physical objects in the games [Caesar III]."
Then, he added, "I know that in the real world the answer is more likely how close you are to the senators, themselves—that defines class. But it's kinda the same."[46]

Colin's story illustrates two important aspects of simulations for learning. First, students often find simulations far more compelling than more traditional ways of representing knowledge; consequently, they spend more time engaging with them and make more discoveries. Second, students experience what they have learned from a robust simulation as their own discoveries. These simulations expose players to powerful new ways of seeing the world and encourage them to engage in a process of modeling, which is central to the way modern science operates. Many contemporary games—*Railroad Tycoon,* for instance— incorporate spreadsheets, maps, graphs, and charts that students must learn to use to play the game. Students are thus motivated to move back and forth across this complex and integrated information system, acting on the simulated environment on the basis of information gleaned from a wide range of different representations.

As games researcher Eric Klopfer cautions, however, simulations enhance learning only when we understand how to read them:

As simulations inform us on anything from global warming to hurricane paths to homeland security, we must know how to interpret this information. If we know that simulations give us information on probabilities we can make better decisions. If we understand the assumptions that go into simulations we can better evaluate that evidence and act accordingly. Of course this applies to decision makers who must act upon that information (police, government, insurance, etc.); it also is important that each citizen should be able to make appropriate decisions themselves based on that information. As it is now, such data is either interpreted by the general public as "fact" or on the contrary "contrived data with an agenda." Neither of these perspectives is useful and instead some ability to analyze and weigh such evidence is critical. Simulations are only as good as their underlying models. In a world of competing simulations, we need to know how to critically assess the reliability and credibility of different models for representing the world around us.[47]

Students who use simulations in learning have more flexibility from customizing models and manipulating data to exploring questions that have captured their own curiosity. There is a thin line between reading a simulation (which may involve changing variables and testing outcomes) and designing simulations. As new modeling technologies become more widely available and as the toolkits needed to construct such models are simplified, students have the opportunity to construct their own simulations. Ian Bogost argues that computer games foster what he calls *procedural literacy*, a capacity to restructure and reconfigure knowledge to look at problems from multiple vantage points, and through this process to develop a greater systemic understanding of the rules and procedures that shape our everyday experiences. He writes, "Engendering true procedural literacy means creating multiple opportunities for learners—children and adults—to understand and experiment with reconfigurations of basic building blocks of all kinds."[48] Young people are learning how to work with simulations through their game play, and schools should build on such knowledge to help them become critical readers and effective designers of simulation and modeling tools. They need to develop a critical vocabulary for understanding the kind of thought experiments performed in simulations and the way these new digital resources inform research across a range of disciplines.

What Might Be Done
Students need to learn how to manipulate and interpret existing simulations and how to construct their own dynamic models of real-world processes.

• Teachers in business classes often ask students to make imaginary investments in the stock market and then monitor actual business reports to track the rise and fall of their "holdings." This well-established classroom practice mirrors what youths do when they form fantasy sports leagues, tracking the performance of players on the sports page to score their results and engaging in imaginary trades to enhance their team's overall standings. Both of these practices share movement between imaginary scenarios (pretend investments or teams) and real-world data. The simulated activities introduce them to the logic by which their real-world counterparts operate and to actual data sets, research processes, and information sources.

• Groups such as OnRampArts in Los Angeles, Urban Games Academy in Baltimore and Atlanta, or Global Kids in New York City involve kids in the design of their own games. These groups see a value in having youths translate a body of knowledge—the history of the settlement of the New World, in the case of OnRampArts' Tropical America—into the activities and iconography of games. Here, students are encouraged to think of alternative ways of modeling knowledge, and they learn to use the vocabulary of game design to represent central aspects of the world around them.

• Simulation games such as *SimCity* provide a context for learning a skill Andy Clark calls *embracing co-control*.[49] In *SimCity*, creating and maintaining a city requires exerting various forms of indirect control. Instead of having top-down control to design a happy, thriving city, the player must engage in a bottom-up process, where the player "grows" a city by manipulating such variables as zoning and land prices. It is only through gaining a familiarity with all the parts of the system, and how they interact, that the player is able to nudge the flow in a way that respects the flow. Such a skill can be understood as a process of "com[ing] to grips with decentralized emergent order,"[50] a mandatory skill for understanding complex systems.

• Students in New Mexico facing a summer of raging forest fires throughout their home state used simulations to understand how flames spread. Manipulating factors such as density of trees, wind, and rain, they saw how even minute changes to the environmental conditions could have profound effects on fire growth. This helped them understand the efficacy of common techniques such as forest thinning and controlled burns.

Performance: The Ability to Adopt Alternative Identities for the Purpose of Improvisation and Discovery

We have thus far focused on game play as a mode of problem solving that involves modeling the world and acting on those models. Yet, game play also is one of a range of contemporary forms of youth popular culture that encourage young people to assume fictive identities and, through this process, develop a richer understanding of themselves and their social roles. In *What Video Games Have to Teach Us about Learning and Literacy*, Gee coins the term *projective identity* to refer to the fusion that occurs between game players and their avatars, the personas they assume in the game. Gee sees the term as playing on two senses of the word *project*: "to project one's values and desires onto the virtual character" and "seeing the virtual character as one's own project in the making."[51] This projected identity allows the player to strongly identify with the character and thus have an immersive experience within the game, and at the same time to use the character as a mirror to reflect on his or her own values and choices.

Testing the educational video game *Revolution* with middle-school students, Russell Francis found several compelling examples in which projected identities had pedagogical payoffs for participants. For example, Margaret played a loyalist character in the game, which is set in colonial Williamsburg on the eve of the American Revolution, and was shaken when she was shot by the redcoats in the midst of a street riot:

The townspeople were very mad. They went to the Governor's mansion to attack. I support the red coats, but they started shooting at me, and then they arrested me. I felt horrified that they would do something like that to me. I don't even believe in violence. I wonder what is going to happen to me. I run the tavern and I have no family. Will I get sent back to England or will I be able to stay here?[52]

She had seen herself as a supporter of the British troops, and at worst an innocent bystander, but she came away from the experience with critical insights about political violence.

Francis built on this process of introspection and projection by asking students to write journals or compose short films reflecting in character on the events that unfolded in the game. In constructing and inhabiting these virtual characters, participants drew together multiple sources of knowledge, mixing things they had read or learned in other educational contexts, information explicitly contained within the game, and their own introspection based on life experiences to create characters that were more compelling to them than the simple digital avatars the designers had constructed. The process closely parallels what actors do when preparing for a role. Here, for example, is how a young African-American girl explained her experiences in playing Hannah, a house slave (an explanation that reaches well beyond anything explicitly present in the game; she even invents actions for the nonplayer characters in order to help her make sense of her place in the social order being depicted):

You don't really have as much support as you would like because being a house slave they call you names, just because most of the time you're lighter skin—you're the master's kid technically. . . . I had to find the ways to get by because, you know, it was hard. On one side, you don't want to get on the Master's bad side because he can beat you. On the other side, the slaves, they ridicule you and are being mean.[53]

Children acquire basic literacies and competencies by learning to manipulate core cultural materials. In *The Braid of Literature: Children's World of Reading*, Wolf and Heath trace the forms of play that shaped Wolf's two preschool-aged daughters' relationship to the "world of words" and stories. Wolf and Heath are interested in how children embody the characters, situations, generic rules, and even specific turns of phrase through their sociolinguistic play. Children do not simply read books or listen to stories; they reenact these narratives in ways that transform them and in this process, the authors argue, children demonstrate that they understand what they have read. This play helps them to navigate the world of stories and, at the same time, elements of stories help them to navigate real-world social situations. Children learn to verbalize their experiences of reading through these performances, and in the process they develop an analytic framework for thinking about literacy.[54]

Anne Haas Dyson's *Writing Superheroes: Contemporary Childhood, Popular Culture, and Classroom Literacy* extends this analysis of the connection between performance and literacy into the classroom, exploring how educators have used dramatizations to teach children to reflect more deeply on their experiences of stories.[55] Wolf and Heath describe individualized play in the context of the home; Dyson recounts social play among peers. In both cases, children start with a shared frame of reference—

stories they have in common, genres they all understand—to ensure that they understand the roles they are to play and the rules of their interaction. Performing these shared fantasies (such as the scenarios that emerge in superhero comics) allows children to better understand who they are and how they connect with the other people around them.

Role-play is very popular with contemporary youths, whether it be the cosplay of young anime fans (costume play based on characters from anime), the fusion with a digital avatar in computer gaming or fantasy role-play, or the construction of alternative personas in subculture communities such as the Goths. Such play has long been understood as testing identities, trying on possible selves, and exploring different social spaces. Susannah Stern stresses the forms of self-representation that are evident on teenagers' Web sites and blogs: "The ability to repeatedly reinvent oneself is particularly appealing since home pages and blogs can be updated as often as desired and because they may be produced anonymously."[56]

These more elaborate and complex forms of role-play may also provide a point of entry into larger spheres of knowledge. Consider, for example, this interview with a seventeen-year-old American girl:

I have been really interested in Japanese culture since I was in sixth grade. When I was in the seventh grade, I started studying Japanese on my own. When I got into high school, I started taking Japanese courses at Smith College. I got into costuming through anime, which is actually how I got interested in Japanese. And I taught myself how to sew. . . . I'm a stage hog. I like to get attention and recognition. I love acting and theater. The biggest payoff of cos-play [costume play] is to go to the conventions where there are other people who know who you are dressed as and can appreciate your effort. At the first convention I ever went to, I must have had fifty people take my picture and at least ten of them came up and hugged me. It's almost like whoever you dress up as, you become that person for a day. . . . People put the pictures up on their websites after the con. So after a con, you can search for pictures of yourself and if you are lucky, you will find five or ten.[57]

For this young girl, assuming the role of a Jpop (Japanese pop) character demonstrates her mastery over favorite texts. Assuming this new identity requires a close analysis of the originating texts, genre conventions, social roles, and linguistic codes. She must go deep inside the story to find her own place within its world. In this case, she also has to step outside the culture that immediately surrounds her to embrace a text from a radically different cultural tradition. She has sought out more information about forms of Asian popular culture. In the process, she has begun to reimagine her relations to the world—as part of an international fan culture that remains deeply rooted in the everyday life of Japan. This search for more information expresses itself across a range of media: the videos or DVDs she watches of Japanese anime, the MP3 or CD recordings of Jpop music, information on the Internet and information she shares with her fellow fans about her own activities, the costumes she generates as well as the photographs of her costumes, the magazines and comics she reads to learn more about Japanese popular culture, and her face-to-face contacts with fellow fans. These activities that center on popular culture in turn translate into other types of learning. As a middle-school student, she began to study Japanese language and culture first on her own and later at a local college.

Role-play should be seen as a fundamental skill used across multiple academic domains. Whether it be children on a playground acting out and deciphering the complex universe of Pokémon, Orville Wright pretending to be a buzzard gliding over sand dunes, or Albert Einstein imagining himself to be a photon speeding over the earth, role-playing enables us to envision and collaboratively theorize about manipulating entirely new worlds.

Role-play, in particular, should be seen as a fundamental skill used across multiple academic domains. So far, we have suggested its relevance to history, language arts, and cultural geography. Yet, this only scratches the surface. Whether it be children on a playground acting out and deciphering the complex universe of Pokémon, Orville Wright pretending to be a buzzard gliding over sand dunes, or Albert Einstein imagining himself to be a photon speeding over the earth, role-playing enables us to envision and collaboratively theorize about manipulating entirely new worlds. Consider, for example, the way role-play informs contemporary design processes. Increasingly, designers construct personae of would-be users, which can serve to illustrate different contexts of use or different interests in the product. These personae are then inserted into fictional scenarios, allowing designers to test the viability of a design and its ability to serve diverse needs. In some cases, this process also involves the designers themselves acting out the different roles and thereby identifying the strengths and limits of their approaches. Improvisational performance, then, represents an important life skill, one that balances problem solving and creative expression, invites a reimagining of self and world, and allows participants to examine a problem from multiple perspectives.

Educators have for too long treated role-play as a means to an end, a fun way to introduce other kinds of content, but we argue that role-play skills may be valuable in their own right and are increasingly central to the way adult institutions function. Performance brings with it capacities to understand problems from multiple viewpoints, to assimilate information, to exert mastery over core cultural materials, and to improvise in response to a changing environment. As with play and simulation, performance places a new stress on learning processes, more on how we learn rather than on what we learn. These learning processes are likely to sustain growth and learning well beyond the school years.

What Might Be Done
Performance enters into education when students are asked to adopt fictive identities and think through scenarios from those characters' perspectives. These identities may be assumed within the physical world or in the virtual world.

• The Model United Nations, a well-established educational project, brings together students from many different schools, each representing delegations from different member countries. Over the course of a weekend, participants work through current debates in foreign policy and simulate the actual procedures and policies of the international organization. Students prepare for the Model United Nations by conducting library research, listening to lectures, and participating in group discussions; they return from the event to share what they learned with other classmates through presentations and written reports.

- The Savannah Project, created by researchers at the University of Bristol, encourages children to play the parts of lions stalking their prey. As they navigate physical spaces, such as the school playground, they receive fictional data via handheld devices. This approach encourages students to master the complex ecosystem of the veldt from the inside out—learning the conditions that impinge on the lion's chances of survival and the skills they need to feed on other local wildlife.

- Teachers in a range of subjects can deploy what David Shaffer calls *epistemic games*.[58] In an epistemic game, the game world is designed to simulate the social context of a profession (say, urban planning), and by working through realistic but simulated problems, players learn the ways of acting, interacting, and interpreting that are necessary for participating in the professional community. In effect, rather than memorizing facts or formulas, through performances of being an urban planner, lawyer, doctor, engineer, carpenter, historian, teacher, or physicist, the player learns these professions' particular ways of thinking.

- Medieval Space, a MySpace clone created by teachers at Byrd Middle School, asked students to create online profiles for the various historical figures studied in their classes. Rather than seeing figures such as Richard III, Henry VI, and Queen Elizabeth as distinct characters, students explored the complex social relationships among them by imagining how they might have interacted had they had online spaces in the fifteenth and sixteenth centuries. For example, students were asked to imagine what their character's current song might be, with 2Pac's "Only God Can Judge Me Now" listed for Richard III.[59]

Appropriation: The Ability to Meaningfully Sample and Remix Media Content

Journalists have frequently used *Napster generation* to describe the young people who have come of age in this era of participatory culture, reducing their complex forms of appropriation and transformation into the simple—arguably illegal—action of ripping and burning someone else's music and sharing the files. Recall that the Pew study found that almost one-quarter of American teens had sampled and remixed existing media content (music, film clips, images, etc.).[60] The digital remixing of media content makes visible the degree to which all cultural expression builds on what has come before it. Appropriation is understood here as a process by which students learn by taking culture apart and putting it back together.

Art does not emerge out of whole cloth from individual imaginations. Rather, it emerges through the artist's engagement with previous cultural materials. Artists build on, are inspired by, appropriate, and transform other artists' work. They do so by tapping into a cultural tradition or by deploying the conventions of a particular genre. Beginning artists often undergo an apprenticeship, during which they try on for size the styles and techniques of other, more established artists. Even well-established artists work with images and themes that have some currency within the culture. Of course, this is not how we generally

---

Appropriation is understood here as a process by which students learn by taking culture apart and putting it back together.

Homer remixed Greek myths to construct *The Iliad* and *The Odyssey*; Shakespeare sampled his plots and characters from other author's plays; the Sistine Chapel ceiling mashes up stories and images from across the entire biblical tradition.

talk about creativity in schools, where the tendency is to discuss artists as individuals who rise above or stand outside any aesthetic tradition. All artists work within traditions; they all also violate conventions. School discourse, however, focuses on one over the other.

Our focus on autonomous creative expression falsifies the actual process by which meaning is generated and new works produced. Most of the classics we teach in the schools are themselves the product of appropriation and transformation, or what we would now call *sampling* and *remixing*. Homer remixed Greek myths to construct *The Iliad* and *The Odyssey*; Shakespeare sampled his plots and characters from other author's plays; the Sistine Chapel ceiling mashes up stories and images from across the entire biblical tradition. Lewis Carroll spoofs the vocabulary of exemplary verses that were then standard to formal education. Many core works of the Western canon emerged through a process of retelling and elaboration: the figure of King Arthur shifts from an obscure footnote in an early chronicle to the full-blown character in *Le Morte d'Arthur* within a few centuries, as the original story is built on by many generations of storytellers.

Many of the forms of expression that are most important to American youths accent this sampling and remixing process, in part because digitization makes it much easier to combine and repurpose media content than ever before. Jazz, for example, evolved through improvisation around familiar themes and standard songs, yet the digital remixing of actual sounds that occurs in techno or hip-hop music has raised much greater alarm among those who would insist on strong copyright protection. Fan fiction (stories about characters or settings in original works written by fans of the original work, not by the original authors) clearly involves the transformative use of existing media content, yet it is often treated as if it were simply a new form of piracy. Collage has been a central artistic practice throughout the twentieth century, one closely associated with the kinds of new creative works that youths are generating by manipulating images with software such as Photoshop. Despite the pervasiveness of these cultural practices, arts and creative-writing programs in schools remain hostile to overt signs of repurposed content, emphasizing the ideal of the autonomous artist. In doing so, they sacrifice the opportunity to help youths think more deeply about the ethical and legal implications of repurposing existing media content, and they often fail to provide the conceptual tools students need to analyze and interpret works produced in this appropriative process.

Appropriation may be understood as a process that involves both analysis and commentary. Sampling intelligently from the

---

Arts and creative-writing programs in schools remain hostile to overt signs of repurposed content, emphasizing the ideal of the autonomous artist. In doing so, they sacrifice the opportunity to help youths think more deeply about the ethical and legal implications of repurposing existing media content.

existing cultural reservoir requires a close analysis of the existing structures and uses of this material; remixing requires an
appreciation of emerging structures and latent potential meanings. Often remixing involves the creative juxtaposition of
materials that otherwise occupy very different cultural niches.
For beginning creators, appropriation provides a scaffolding,
allowing them to focus on some dimensions of cultural production and rely on the existing materials to sustain others. They
are able, for example, to focus more attention on description or
exposition if they can build on existing characters and plots.
They learn how to capture the voice of a character by trying to
mix borrowed dialog with their own words. Mapping their emotional issues onto preexisting characters allows young writers to
reflect on their own lives from a certain critical distance and
work through issues, such as their emerging sexualities, without
facing the stigma that might surround confessing such feelings
in an autobiographical essay. These students learn to use small
details in the original works as probes for their own imagination, overcoming some of the anxiety of staring at a blank computer screen. Building on existing stories attracts wider interest
in their work, allowing it to circulate far beyond the community
of family and friends. In turn, because they are working with a
shared narrative and many others have a stake in what happens
to these characters, they receive more feedback on their writing.

> Sampling intelligently from the existing cultural reservoir requires
> a close analysis of the existing structures and uses of this material;
> remixing requires an appreciation of emerging structures and
> latent potential meanings.

What Might Be Done

Appropriation enters education when learners are encouraged to dissect, transform, sample, or remix existing cultural materials.

- The MIT Comparative Media Studies Program hosts a workshop each year asking students to work in teams to think through what would be involved in transforming an existing media property (a book, film, television series, or comic book) into a video or computer game and then preparing a "pitch" presentation for their game. Beginning with a preexisting property allows students to start quickly and more or less on equal footing because they are able to build on a text they have in common as readers rather than one created by an individual student author. The process of identifying core properties of the original work teaches students important skills in narrative and formal analysis, while the development of an alternative version of the story in another medium emphasizes the creative expansion of the original content.[61]
- The crew of Public Radio International's program *Sound & Spirit* has encouraged students in the greater Boston area to develop scripts and record radio broadcasts that use critical commentary of existing songs to explore a common theme or topic. They have found that this process of sampling and remixing music motivates youths to think more deeply about the sounds they hear around them and motivates them to approach school-related topics from a fresh perspective.
- Artist and filmmaker Juan Devis has been working with the University of Southern California Film School, the Institute for Media Literacy, and the Los Angeles Leadership Academy on a project with minority youths. The youths will develop an online

game based on Mark Twain's *Huckleberry Finn*. Devis drew a number of strong parallels between the experiences of minority youths in Los Angeles and the world depicted in Twain's novel—including parallels between "crews" of taggers and the gang of youths that surrounds Huck and Tom, the use of slang as a means of separating themselves from their parents' culture, the complex experience of race in a society undergoing social transitions, and the sense of mobility and "escape" from adult supervision.[62]

- Ricardo Pitts-Wiley, the artistic director of Mixed Magic Theatre, has been working with students from high schools in Pawtucket, Rhode Island, to develop his "urban Moby Dick" project. Students have worked closely with mentors—artists, law enforcement officers, and business leaders from the local community—to explore Herman Melville's classic novel together. Through a process of reading, discussion, improvisation, and writing, they are scripting and staging a modern version of the classic whaling story, one that acknowledges the realities of contemporary urban America. In their version, the "Great White" turns out not to be a whale but an international drug cartel. Ish and Quay are two members of Ahab's posse as he goes after the vicious force that took his leg and killed his wife. Through reimagining and reworking Melville's story, they have come to a deeper understanding of the relationships between the characters and of some of the core themes of male bonding and obsession that run through the book.[63]

- Renee Hobbs, a twenty-year veteran of the media literacy movement, recently launched a new Web site—*My Pop Studio*—that encourages middle-school and early-high-school girls to reflect more deeply about some of the media they consume—

pop music, reality television, celebrity magazines—by stepping into the roles of media producers. The site offers a range of engaging activities, including designing an animated pop star and scripting their next sensation, reediting footage for a reality television show, and designing the layout for a teen magazine. They are asked to reflect on the messages the media offers on what it is like to be a teen girl in America today and to think about the economic factors shaping the culture that has become so much a part of their everyday interactions with their friends.[64]

Multitasking: The Ability to Scan the Environment and Shift Focus onto Salient Details

Perhaps one of the most alarming changes, in adults' view, is the perceived decline of young people's attention span with the rise of digital media. Attention is undoubtedly an important cognitive ability. All information to be processed by our brains is temporarily held in short-term memory, and the capacity of our short-term memory is sharply limited.[65] Attention is critical. Learners must filter out extraneous information and sharpen their focus on the most salient details of their environment. Instead of focusing on narrowing attention, young people often respond to a rich media environment by multitasking—scanning for relevant shifts in the information flow while simultaneously taking in multiple stimuli. Multitasking and attention should not be seen as oppositional forces. Rather, we should think of them as two complementary skills, both strategically employed by the brain to intelligently manage constraints on short-term memory. Whereas attention seeks to prevent information overload by controlling what information enters short-term memory, successful multitaskers seek to reduce demands on short-term memory by mapping where different information is externally stored within their immediate environment.

In "Growing up Digital," John Seely Brown describes an encounter he had:

Recently I was with a young twenty-something who had actually wired a Web browser into his eyeglasses. As he talked with me, he had his left hand in his pocket to cord in keystrokes to bring up my Web page and read about me, all the while carrying on with his part of the conversation! I was astonished that he could do all this in parallel and so unobtrusively. . . . People my age tend to think that youths who are multiprocessing can't be concentrating. That may not be true. Indeed, one of the things we noticed is that the attention span of the teens at PARC—often between 30 seconds and five minutes—parallels that of top managers, who operate in a world of fast context-switching. So the short attention spans of today's youth may turn out to be far from dysfunctional for future work worlds.[66]

Currently, young people are playing with these skills as they engage with games or social activities that reward the ability to maintain a mental picture of complex sets of relationships and to adjust quickly to shifts in perceptual cues. The multitasking process is already evident in the "scrawl" on television news: the screen is a series of information surfaces, each containing a relevant bit of data, none of which offers the complete picture.[67] Our eyes scan across electoral maps and ticker tapes, moving images and headlines, trying to complete a coherent picture of the day's events and to understand the relationship between the visuals. Similarly, as Gunther Kress notes, the contemporary textbook increasingly deploys a broader array of different modalities as it represents information students need to know about a given topic.[68] Here, again, readers are being taught to scan the informational environment rather than fix attention on a single element.

Historically, we might have distinguished between the skills required of farmers and those expected of hunters. The farmer must complete a sequence of tasks that require localized attention; the hunter must scan a complex landscape in search of signs and cues of where prey may be hiding. For centuries, schools have been designed to create "farmers."[69] In such an organization, the ideal is for all students to focus on one thing; indeed, attention is conceived of as the ability to concentrate on one thing for an extended period of time, while the inability or refusal to maintain such a narrow focus is characterized as a "disorder." Yet, fixed attention would be maladjusted to the needs of hunters, who must search high and low for their game. Schools adapted to the needs of "hunters" would have very different practices and might well value the ability to identify the relationship between seemingly unrelated developments within a complex visual field. As we look to the future, one possibility is that schools will be designed to support both styles, ensuring that each child develops multiple modes of learning and multiple strategies for processing information. In such a world, neither attentional style is viewed as superior, but both are assessed in terms of their relative value within a given context.

Multitasking often is confused with distraction, but, as understood here, multitasking involves a method of monitoring and responding to the sea of information around us. Students need help distinguishing between being off task and handling multiple tasks simultaneously. They must learn to recognize the rela-

Multitasking often is confused with distraction, but, as understood here, multitasking involves a method of monitoring and responding to the sea of information around us. Students need help distinguishing between being off task and handling multiple tasks simultaneously.

tionship between information coming at them from multiple directions and making reasonable hypotheses and models based on partial, fragmented, or intermittent information (all part of the world they will confront in the workplace). They need to know when and how to pay close attention to a specific input as well as when and how to scan the environment to search for meaningful data.

What Might Be Done

Multitasking enters pedagogical practice when teachers recognize the desires of contemporary students to come at topics from multiple directions all at the same time or to maintain what some have called "continuous partial attention," interacting with homework materials while engaged in other activities.

• A teacher's assistant blogs in real time in response to the classroom instructor's lectures, directing students' attentions to relevant links that illustrate and enhance the content being discussed, rather than providing distractions from the core activity. Students are encouraged to draw on this related material as they engage in classroom discussion, grounding their comments in specific examples and quotations from relevant documents.

• At the Brearley School in Manhattan, foreign-language class materials are transferred directly from the school's computer servers to students' iPods. Rather than needing to set aside dedicated study time to practice a foreign language, students can access their homework and foreign songs while walking home from classes or engaging in other activities.[70]

• The online game *Cyber Nations*, a simulation that helps players learn about nation building and international diplomacy, breaks player actions down into distinct choices that can be made at the player's own pace. This encourages players to keep a browser window open to periodically check for updates from their nation throughout the day while working on other tasks, rather than playing the game only during a dedicated play time. Homework assignments in the form of online games could be designed in a similar manner to facilitate patterns of multitasking.

Distributed Cognition: The Ability to Interact Meaningfully with Tools That Expand Mental Capacities

Challenging the traditional view that intelligence is an attribute of individuals, the distributed cognition perspective holds that intelligence is distributed across "brain, body, and world,"[71] looping through an extended technological and sociocultural environment.[72] Explaining this idea, Roy Pea notes, "When I say that intelligence is distributed, I mean that the resources that shape and enable activity are distributed in configurations across people, environments, and situations. In other words, intelligence is accomplished rather than possessed."[73] Work in distributed cognition focuses on forms of reasoning that would not be possible without the presence of artifacts or information appliances and that expand and augment human cognitive capacities. These devices might be forms that externalize memory, such as a database, or they can be devices that externalize processes, such as the widely used spell-checker.[74] The more we rely on the capacities of technologies as a part of our work, the more it may seem that cognition is distributed.

Teachers have long encouraged students to bring scratch paper with them into math examinations, realizing that the ability to construct representations and record processes is vital in solving complex problems. If, as Clark notes, technologies are inextricably interwoven with thinking, it makes no sense to "factor out" what the human brain is doing as the "real" part of thinking and to view what the technology is doing as a "cheat" or "crutch."[75] Rather, we can understand cognitive activity as shared among a number of people and artifacts, and cognitive acts as learning to think with other people and artifacts. Following this theory, students need to know how to think with and through their tools as much as they need to record information in their heads.

Gamers may be acquiring some of these distributed cognition skills through their participation in squadron-based video games. Gee suggests that playing such games requires a mental map of what player and nonplayer characters are doing (nonplayer characters are characters controlled by the artificial intelligence of the game).[76] To plan appropriately, players may not need to know what other participants know, but they do need to know what it is those participants are likely to do. Moreover, in playing the games, a player may need to flip through a range of different representations of the state of the game world and of the actions that are occurring within it. Learning to play involves learning to navigate this information environment, understanding the value of each representational technology, and knowing when to consult each and how to deploy this knowledge to reshape what is occurring. Instead of thinking as an autonomous problem solver, the player becomes part of a social and technological system that is generating and deploying information at a rapid pace. Humans are able to play much more complex games (and to solve much more complex problems) in a world in which keeping track of key data and enacting well-understood computational processes can be trusted to the processing power of the computer; the human can thus focus more attention on strategic decision making.

Distributed cognition is not simply about technologies; it is also about tapping social institutions and practices or remote experts whose knowledge may be useful in solving a particular problem. According to this understanding, expertise comes in many shapes and sizes (both human and nonhuman). Experts can be expert practitioners, who can be consulted through such technologies as video conferencing, instant messaging, or email;

Distributed intelligence is not simply a technical skill. Rather, it is a cognitive skill that involves thinking across "brain, body, and world." The term *distributed intelligence* emphasizes the role that technologies play in this process, but it is closely related to the social production of knowledge that we are calling *collective intelligence*.

> Students need to learn the affordance of different tools and information technologies, to know what the technologies can do and in what contexts they can be trusted. Students need to acquire patterns of thought that regularly cycle through available sources of information as they make sense of developments in the world around them.

some knowledge can emerge from technologies such as calculators, spread sheets, and expert systems; and new insights can originate from the teacher or students or both. The key is having expertise somewhere within the distributed learning environment and making sure students understand how to access and deploy it.

Applications of the distributed cognition perspective to education suggest that students must learn the affordances of different tools and information technologies and know which functions tools and technologies excel at and in what contexts they can be trusted. Students need to acquire patterns of thought that regularly cycle through available sources of information as they make sense of developments in the world around them. Distributed intelligence is not simply a technical skill, although it depends on knowing how to use tools effectively; it is also a cognitive skill that involves thinking across "brain, body, and world." The term *distributed intelligence* emphasizes the role that technologies play in this process, but it is closely related to the social production of knowledge that we are calling *collective intelligence*.

What Might Be Done

The theory of distributed cognition informs educational research and practice when it provides a perspective for envisioning new learning contexts, tools, curricula and pedagogy, participant structures, and goals for schooling.

▪ Augmented-reality games represent one potential application of distributed intelligence to the learning process. Klopfer and Squire developed a range of games in which students use location-aware, GPS-enabled handheld computers to solve fictional problems in real spaces. For example, in *Environmental Detectives*, students determine the source of an imaginary chemical leak that is causing environmental hazards on the MIT campus. Students can use their handhelds to drill imaginary wells and take readings on the soil conditions, but to do so they must travel to the actual location. Data drawn from the computer is read against their actual physical surroundings—the distance between locations, the slope of the land, its proximity to the Charles River—and multiple players compare notes as they seek to resolve the game scenario.[77]

▪ Students in MIT's Comparative Media Studies Program have experimented with the use of handhelds to allow tourists to access old photographs of historic neighborhoods and compare them with what they are seeing on location.[78] Elsewhere, students travel across the battlefield at Lexington conducting interviews with historical personages to better understand their perspective on what happened there in 1775.[79] In each case, direct perceptions of the real world and information drawn from information appliances are mutually reinforcing. The players combine multiple information sources in completing the tasks at hand.

• *Byline* is an Internet-based publishing and editing tool designed to focus attention on the organizational and structural features of journalism. By providing a space for the body of the story, the byline, and the lead, this "smart tool" is scaffolding for students' learning process as they write a journalistic story. By cueing students into what to write, where to write it, and even into such journalistic values as the need to catch the reader's attention, this specially designed program helps students to learn the conventions and values of journalism.[80]

• A classroom designed to foster distributed cognition encourages students to participate with a range of people, artifacts, and devices. The various forms of participation composing such cognitive activity might be understood more generally as the skill of knowing how to act within distributed knowledge systems. Interested in designing learning environments that would foster such a skill, Bell and Winn describe a classroom in which participation requires active collaborations not only with people and tools that are physically present but also with people and tools that are virtually present through, for example, video conferencing with a science practitioner, using the Web to connect to a database in Japan, and using Excel spreadsheets to simulate a mass spectrometer. In such classrooms, knowing how to act within the distributed knowledge system is more important than learning content. Because content is something that can be "held" by technologies—databases, Web sites, wikis, and so forth—the curricular focus is on learning how to generate, evaluate, interpret, and deploy data.[81]

• With new technologies, new cognitive possibilities arise. Educators need to create new activities when new technologies are introduced into the classroom. If the calculator is used to add 2+2, it is the capacities of the calculator that are solving the problem; when calculation is "off-loaded" onto the calculator, the student is free to solve more complex problems.[82] The proliferation of digital technologies requires a concerted effort to envision activities that enable students to engage in more complex problem domains. For example, as a vehicle for assessing the various ways e-commerce affects the environment, students could be given the problem of comparing the environmental impact of shipping 250,000 copies of *Harry Potter and the Goblet of Fire* directly to individual customers rather than to bookstores. Reflecting on the intended outcome for such a comparison, Robert Yagelski notes, "The click of the computer mouse to order a copy of *Harry Potter* from Amazon.com can seem a simple and almost natural act, yet it represents participation in this bewilderingly complex web of material connections that is anything but simple. And that participation contributes to the condition of our planet."[83]

Collective Intelligence: The Ability to Pool Knowledge and Compare Notes with Others toward a Common Goal

As users learn to exploit the potential of networked communication, they participate in a process that Pierre Lévy calls *collective intelligence*. Like-minded individuals gather online to embrace common enterprises, which often involve accessing and processing information. In such a world, Lévy argues, everyone knows something, nobody knows everything, and what any one person knows can be tapped by the group as a whole. We are still experimenting with how to work within these knowledge cultures and what they can accomplish when we pool knowledge. Lévy argues that as a society we are currently in an apprenticeship phase, during which we try out and refine skills and institutions that will sustain the social production of knowledge. Lévy sees collective intelligence as an alternative source of power, one that allows grassroots communities to respond effectively to government institutions that emerge from the nation-state or to corporate interests that sustain multinational commerce.[84] Already we are seeing governments and industries seek ways to "harness collective intelligence," which has become the driving force behind what people are calling Web 2.0.

Currently, children and adults are acquiring the skills to operate within knowledge communities in order to interact with popular culture. As has often been the case, what we learn through play we later apply to more serious tasks. So, for example, the young Pokémon fans, who each know some crucial detail about the various species, constitute a collective intelligence whose knowledge is extended each time two youths on the playground share something about the franchise.

Such knowledge sharing can assume more sophisticated functions as it moves online. For example, *Matrix* fans have created

---

Collective intelligence . . . in such a world, everyone knows something, nobody knows everything, and what any one person knows can be tapped by the group as a whole.

elaborate guides to help them track information about the fictional Zion resistance movement featured in the film. Young people are playing with collective intelligence as they participate in the vast knowledge communities that emerged from the online game *I Love Bees*. Some estimate that as many as 3 million players participated in *I Love Bees*, history's most challenging scavenger hunt. After working through puzzles so complicated that they mandated the effective collaboration of massive numbers of people with expertise across a variety of domains and geographic locations, players gathered clues by answering more than 40,000 pay-phone calls across all fifty U.S. states and eight countries.[85] They then fed those clues back into online tools designed to support large-scale collaboration for all players to deconstruct and analyze. If players were unfamiliar with how to participate in the community, other players would train them in the necessary skills. In another example, fans of the television show *Survivor* have used the Internet to track down information and identify the names of contestants before they were announced by the network. Fans also have used satellite photographs to identify the location of the Survivor base camp despite the producer's "no flyover" agreements with local governments. These knowledge communities change the very nature of media consumption—a shift from the personalized media that was central to the idea of the digital revolution toward socialized or communalized media that is central to the culture of media convergence.[86]

As players learn to work and play in such knowledge cultures, they come to think of problem solving as an exercise in teamwork. Consider the following postings made by members of the Cloudmakers, a team formed in a game similar to *I Love Bees*:

> These knowledge communities change the very nature of media consumption—a shift from the personalized media that was central to the idea of the digital revolution toward socialized or communalized media that is central to the culture of media convergence.

The solutions do not lie in the puzzles we are presented with, they lie in the connections we make, between the ideas and between one another. These are what will last. I look down at myself and see that I, too, have been incorporated into the whole, connections flowing to me and from me, ideas flowing freely as we work together, as individuals and as a group, to solve the challenges we are presented with. The solution, however, does not lie in the story. We are the solution.

* * *

The 7500+ people in this group . . . we are all one. We have made manifest the idea of an unbelievably intricate intelligence. We are one mind, one voice . . . made of 7500+ neurons. . . . We are not one person secluded from the rest of the world. . . . We have become a part of something greater than ourselves.[87]

Indeed, these groups have been drawn from playing games to confronting real-world social problems, such as tracking campaign finances or trying to solve local crimes, as they develop a new sense of self-confidence in their ability to tackle challenges collectively, challenges that, as individuals, they would be unable to face.

This focus on teamwork and collaboration is also, not coincidentally, how the modern workplace is structured—around ad hoc configurations of employees, brought together because their diverse skills and knowledge are needed to confront a specific challenge and then dispersed into different clusters of workers when new needs arise. Cory Doctorow has called such systems *ad-hocracies*, suggesting that they contrast in every possible way with prior hierarchies and bureaucracies. Our schools do an excellent job, consciously or unconsciously, of teaching youths how to function within bureaucracies. They do almost nothing to help youths learn how to operate within an ad-hocracy.[88]

Collective intelligence is increasingly shaping how we respond to real-world problems. On August 29, 2005, Hurricane Katrina tore apart the levee that protected New Orleans from Lake Pontchartrain and the Mississippi River. Not only was the ability of ordinary citizens to share self-produced media and information pivotal in shaping the view of the situation for the outside world (thereby bringing in more relief funds), but it allowed for those affected by the disaster to effectively assist one another. After Jonathan Mendez's parents evacuated from Louisiana to his home in Austin, Texas, he was eager to find out if the floods had destroyed their home in Louisiana. Unfortunately for him, media coverage of the event was focused exclusively on the most devastated parts of New Orleans, with little information about the neighborhood where his parents had lived. With some help from his coworker, they were able, within a matter of hours, to modify the popular Google Maps Web service to allow users to overlay any information they had about the devastation directly onto a satellite map of New Orleans. Shortly after making their modification public, more than 14,000 submissions covered their map. This allowed victims scattered throughout the United States to find information about any specific location—including verifying that the Mendez's house was still intact.[89]

Unfortunately, most contemporary education focuses on training autonomous problem solvers and is not well suited to equip students with these skills. Collective intelligence communities encourage ownership of work as a group, but schools grade individuals. Although Jonathan Mendez was admired for having appropriated Google's Internet mapping service, students in school are often asked to swear that what they turn is their "own work."

Leadership within a knowledge community requires the ability to identify specific functions for each member of the team based on his or her expertise and to interact with the team members in an appropriate fashion. Teamwork involves a high degree of interdisciplinarity—reconfiguring knowledge across traditional categories of expertise. In early February 2004, Eric Klopfer, an MIT professor of urban studies and planning, and a team of researchers from the Education Arcade conducted "a hi-tech whodunit" for middle-school students and their parents inside the Boston Museum of Science. Teams of three adult-child pairs were given handhelds to search for clues of the whereabouts and identities of the notorious Pink Flamingo Gang, who had stolen an artifact and substituted it with a fake. Thanks to the museum's newly installed Wi-Fi network and the players' location-aware handhelds, each gallery offered the opportunity to interview cyber-suspects, download objects, examine them with virtual equipment, and trade findings. Each parent-child unit was assigned a different role—biologists, detectives, or technologists—enabling them to use different tools on the evidence they gathered.[90] This is simply one of many recent cooperative games assigning distinct roles to each player, giving each access to a different set of information, and thus creating strong incentives for them to pool resources with other players.

Schools, on the other hand, often seek to develop generalists rather than allowing students to assume different roles based on their emerging expertise. The ideal of the Renaissance man is someone who knows everything, or at least a great deal about a wide range of different topics. The ideal of a collective intelligence is a community that knows everything, with individuals who know how to tap the community to acquire knowledge on a just-in-time basis. Minimally, schools should be teaching students to thrive in both worlds: having a broad background on a range of topics but also knowing when to turn to a larger community for relevant expertise. They must know how to solve problems on their own but also how to expand their intellectual capacity by working on a problem within a social community. To be meaningful participants in such a knowledge culture, students must acquire greater skills for assessing the reliability of information, which may come from multiple sources, some of which are governed by traditional gatekeepers and others of which must be cross-checked and vetted within a collective intelligence.

What Might Be Done

Schools can deploy aspects of collective intelligence when students pool observations and work through interpretations with others studying the same problems at scattered locations. Such knowledge communities can confront problems of greater scale and complexity than any given student might be able to handle.

• Scientists in fields requiring simple yet extensive data-analysis tasks could partner with middle-school teachers to have students help collect or analyze real data. Eelgrass is both the most abundant sea grass in Massachusetts and one of the most ecologically valuable marine and estuarine habitats in North American coastal waters. The MIT Sea Grant College Program developed a project where students in different schools learn to cultivate eelgrass and collaboratively share data regarding the levels of nitrates, oxygen, and so forth in affected habitats through the project Web site.[91]

• Sites such as Ning offer nonprogrammers tools for rapidly creating social Web applications that allow users to interact and share information with one another. For example, a Mandarin teacher could easily create an online travel guide in which students (potentially nationwide) would each contribute write-ups of interesting sites in their local areas that would be of interest to visitors from China.

• Students taking civics classes might be encouraged to map their local governments using a Wikipedia-like program, bringing together names of government officials, reports on government meetings, and key policy debates. The information would be accessible to others in their own communities. They might also compare notes with students living in other parts of the country to identify policy alternatives that might address problems or concerns in their communities.

Judgment: The Ability to Evaluate the Reliability and Credibility of Different Information Sources

Although it is exciting to see players harness collective intelligence to successfully solve problems of unprecedented complexity, this process also involves a large number of errors. Misinformation emerges and is worked over, refined, or dismissed before a new consensus emerges. We are taught to think of knowledge as a product, but within a collective intelligence, knowledge is also always in process. As such, understanding the current placement within the vetting process helps determine how much trust to place in any given piece of information. In a game such as *I Love Bees*, mistakes are generally of little consequence and often serve as much as a source of amusement as anything else. As these same technologies are employed in understanding world events, though, we must better understand the strengths and limitations of these new practices of knowledge production.

For example, one key technology in online collective intelligence communities is the wiki. Although it may be possible for a small group of individuals to contribute erroneous information, wiki enthusiasts argue that giving all members of a larger community the ability to correct any mistakes will ultimately lead to more accurate information. In many cases, this concept has proved surprisingly effective. In one study, *Nature* magazine compared the accuracy of articles in Wikipedia, an enormous online encyclopedia constructed entirely through the efforts of volunteers using wiki technologies, with equivalent articles in *Encyclopedia Britannica*. The study determined the accuracy levels of the two to be roughly the same. Wikipedia is not flaw-less, but rather even sources such as *Encyclopedia Britannica* are flawed.[92] Students must be taught to read both sources from a critical perspective.

The *Nature* article also points out that wikis perform best when a large number of participants are actively using the technology to correct mistakes. Whereas the Wikipedia article on global warming enjoys more than 10,000 authors, each passionately committed to ensuring the accuracy of its content, the biographical article on John Seigenthaler cited him as having a possible involvement in the assassinations of Robert F. Kennedy and John F. Kennedy for a period of 132 days before someone corrected it.[93] Given the disparity in the accuracy of different articles, students need to develop an intuitive understanding of how the contents of a wiki are produced by participating in their construction and then actively reflecting on the different possibilities for inaccuracies.

In truth, schools should always teach students critical thinking skills for "sussing out" the quality of information, yet historically schools have had a tendency to fall back on the gatekeeping functions of professional editors and journalists, not to mention those of textbook selection committees and librarians, to ensure that the information is generally reliable. Once students enter cyberspace, where anyone can post anything, they need skills in evaluating the quality of different sources, the probability of perspectives and interests coloring representations, and the likely mechanisms by which misinformation is perpetuated or corrected. We need to balance a trust of traditional gatekeeping organizations (public television,

Smithsonian, National Geographic, for example) with the self-correcting potential of grassroots knowledge communities. Traditional logic would suggest, for example, that *60 Minutes*, a long-standing network news show, would be more accurate than a partisan blog, but in fall 2004, bloggers working together recognized flaws in evidence that had been vetted by the established news agency. As Dan Gillmor notes, we are entering a world in which citizen journalists often challenge and sometimes correct the work of established journalists, even as journalists debunk the urban folklore circulated in the blogging community.[94]

Misinformation abounds online, but so do mechanisms for self-correction. In such a world, we can only trust established institutions so far. We all must learn how to read one source of information against another, to understand the contexts within which information is produced and circulated, to identify the mechanisms that ensure the accuracy of information, and to realize under which circumstances those mechanisms work best. Confronted with a world in which information is unreliable, many of us fall back on cynicism, distrusting everything we read. Rather, we should foster a climate of healthy skepti-

The new mediated landscape of mainstream news sources, collaborative blog projects, unsourced news sites, and increasingly sophisticated marketing techniques aimed at ever-younger consumers demands that students be taught how to distinguish fact from fiction, argument from documentation, real from fake, and marketing from enlightenment.

cism in which all truth claims are weighed carefully, but in which there is an ethical commitment to identifying and reporting the truth.

Students are theoretically taught in school how to critically assess the pros and cons of an argument. In an increasingly pervasive media environment, they also must be able to recognize when arguments are not explicitly identified as such. The new mediated landscape of mainstream news sources, collaborative blog projects, unsourced news sites, and increasingly sophisticated marketing techniques aimed at ever-younger consumers demands that students be taught how to distinguish fact from fiction, argument from documentation, real from fake, and marketing from enlightenment.

"To be a functioning adult in a mediated society, one needs to be able to distinguish between different media forms and know how to ask basic questions about everything we watch, read, or hear," say Share, Jolls, and Thoman. "Although most adults learned through English classes to distinguish a poem from an essay, it is amazing how many people do not understand the difference between a daily newspaper and a supermarket tabloid, what makes one website legitimate and another one a hoax, or how advertisers package products to entice us to buy."[95]

Even when media content has been determined credible, it is vital for students to also identify and analyze the perspective of the producer: who is presenting what to whom, and why. Existing media literacy materials excel in examining the forces behind controversial media properties, particularly provocative visuals, their intentions, and their effects.

As Buckingham notes, children may lack some of the core life experiences and basic knowledge that might help them to discriminate between accurate and inaccurate accounts:

[T]here is as yet relatively little research about how children make judgments about the reliability of information on the Internet, or how they learn to deal with unwelcome or potentially upsetting content. Children may have more experience of these media than many adults, but they mostly lack the real-world experience with which media representations can be compared; and this may make it harder for them to detect inaccuracy and bias.[96]

Reviewing the literature on how children make sense of online resources, Buckingham found that students lack both knowledge and interest in assessing how information was produced for and within digital environments: "Digital content was 'often seen as originating not from people, organisations, and businesses with particular cultural inclinations or objectives, but as a universal repository that simply existed 'out there.'"[97] Other studies found that children remain unaware of the motives behind the creation of Web sites, have difficulty separating commercial from noncommercial sites, and lack the background to identify the sources of authority behind claims made by Web authors.

As this discussion has suggested, judgment might be seen as part of our existing conception of literacy—a core research skill of the kind that has long been fundamental to the school curriculum. Yet this discussion also underscores that judgment operates differently in an era of distributed cognition and collective intelligence. Judgment requires not simply logic but also an understanding of how different media institutions and cultural communities operate. Judgment works not simply on knowledge as the product of traditional expertise but also on the process by which grassroots communities work together to generate and authenticate new information.

What Might Be Done
Judgment has long been the focus of media literacy education in the United States and around the world as students are encouraged to ask critical questions about the information they are consuming.

• The Boston-based Youth Voice Collaborative has developed an exercise that gives students a range of news stories and asks them to rank the stories according to traditional news standards. The process is designed to encourage students to understand what criteria journalists use to determine the "news value" of different events and to encourage students to express their own priorities about what information matters to them and why.

• Google aggregates articles from thousands of news sources worldwide at news.google.com. This allows users to compare and contrast the framing of a single issue from different media sources. Students are encouraged to read several articles closely, underlining words they believe might shape how readers understand and feel about what they are reading.

• The New Media Literacies project at MIT has developed a set of activities to involve students in understanding how representations of "truth" and "fiction" vary in different media forms and, therefore, how different techniques must be learned, and choices made, when seeking to manipulate meanings by altering representations. For example, in an image-manipulation activity, students search for an image of an event (such as the March on Washington or the Kennedy assassination) and are taught how to change the picture in ways that change the meaning. By manipulating images, students become familiar with the ways images may be altered to persuade and influence. In developing this manipulation skill, students are encouraged to think about why an image, sound, or textual representation might be altered and what that means to them as consumers, voters, and citizens.

- A growing number of teachers are using the Talk Pages for contested Wikipedia entries as illustrations of the types of questions to ask about any information and the processes and criteria by which disputes about knowledge might be resolved.

- Tools such as the Ligit site allow readers of a Web site to alert friends who subsequently read the same Web site that its content may be suspect. Students might also be encouraged to take advantage of sites such as Snopes that regularly report on frauds and misinformation circulating online and provide good illustrations of the ways to test the credibility of information.

Transmedia Navigation: The Ability to Follow the Flow of Stories and Information across Multiple Modalities

In an era of convergence, consumers become hunters and gatherers, pulling together information from multiple sources to form a new synthesis.[90] Storytellers exploit this potential for transmedia storytelling, advertisers talk about branding as depending on multiple touch points, and networks seek to exploit their intellectual properties across many different channels. As they do so, we encounter the same information, the same stories, the same characters and worlds across multiple modes of representation. Transmedia stories at the most basic level are stories told across multiple media. At the present time, the most significant stories tend to flow across multiple media platforms.

Consider, for example, the Pokémon phenomenon. As Buckingham and Sefton-Green explain, "Pokémon is something you do, not just something you read or watch or consume." Several hundred different Pokémon exist, each with multiple evolutionary forms and a complex set of rivalries and attachments. There is no one text for information about these various species. Rather, the child assembles information from various media, with the result that each child knows something his or her friends do not. As a result, the child can share his or her expertise with others. As Buckingham and Sefton-Green explain, "Children may watch the television cartoon, for example, as a way of gathering knowledge that they can later utilize in playing the computer game or in trading cards, and vice versa. The fact that information can be transferred between media (or platforms) of course adds to the sense that Pokémon is unavoidable. In order to be a master, it is necessary to 'catch' all its various manifestations."[99]

Such information feeds back into social interactions, including face-to-face contact within local communities and mediated

---

Transmedia stories at the most basic level are stories told across multiple media.

contact online with a more dispersed population. These children's properties offer multiple points of entry, enable many different forms of participation, and facilitate the interests of multiple consumers.[100]

One dimension of this phenomenon points us back to collective intelligence, given that what Ito calls *hypersociability* emerges as children trade notes on and exchange artifacts associated with their favorite television shows. A second dimension of this phenomenon points us to what Kress calls *multimodality*.[101] Consider a simple example: the same character (say, Spider-Man) may look different when featured in an animated video than in a video game, in a printed comic book, as a molded plastic action figure, or in a live-action movie. How then do fans recognize a character across all of these different media? How do they link what they have learned about the character in one context to what they learned in a completely different media channel? How do they determine which of these representations are linked (part of the same interpretation of the character) and which are separate (separate versions of the character that are meant to be understood autonomously)? These are the kinds of conceptual problems youths encounter regularly in their participation in contemporary media franchises.

Kress stresses that modern literacy requires the ability to express ideas across a broad range of different systems of representation and signification (including "words, spoken or written; image, still and moving; musical . . . 3D models . . . "). Each medium has its own affordances, its own systems of representation, its own strategies for producing and organizing knowledge. Participants in the new media landscape learn to navigate these different and sometimes conflicting modes of representation and to make meaningful choices about the best ways to express their ideas in each context.[102] All of this sounds more complicated than it is. As the New Media Consortium's report on twenty-first century literacy suggests, "Young people adept at interpreting meaning in sound, music, still and moving images, and interactive components not only seem quite able to cope with messages that engage several of these pathways at once, but in many cases prefer them."[103]

Kress argues that this tendency toward multimodality changes how we teach composition because students must learn to sort through a range of different possible modes of expression, determine which is most effective in reaching their audience and communicating their message, and grasp which techniques work best in conveying information through this channel. Kress advocates moving beyond teaching written composition to teaching design literacy as the basic expressive competency of the modern era. This shift does not displace printed texts with images, as some advocates of visual literacy have suggested.

> This tendency toward multimodality changes how we teach composition because students must learn to sort through a range of different possible modes of expression, determine which is most effective in reaching their audience, and grasp which techniques work best in conveying information through this channel. This shift does not displace printed texts with images. Rather, it requires students to be equally adept at reading and writing through images, texts, sounds, and simulations.

Rather, it develops a more complex vocabulary for communicating ideas, a vocabulary that requires students to be equally adept at reading and writing through images, texts, sounds, and simulations. The filmmaker George Lucas offers an equally expansive understanding of what literacy might mean today:

We must teach communication comprehensively in all its forms. Today we work with the written or spoken word as the primary form of communication. But we also need to understand the importance of graphics, music, and cinema, which are just as powerful and in some ways more deeply intertwined with young people's culture. We live and work in a visually sophisticated world, so we must be sophisticated in using all the forms of communication, not just the written word.[104]

In short, new media literacies involve the ability to think across media, whether understood at the level of simple recognition (identifying the same content as it is translated across different modes of representation), at the level of narrative logic (understanding the connections among one story communicated through different media), or at the level of rhetoric (learning to express an idea within a single medium or across the media spectrum). Transmedia navigation involves both processing new types of stories and arguments that are emerging within a convergence culture and expressing ideas in ways that exploit the opportunities and affordances represented by the new media landscape. In other words, it involves the ability to both read and write across all available modes of expression.

What Might Be Done

Students learn about multimodality and transmedia navigation when they take time to focus on how stories change as they move across different contexts of production and reception, as they give consideration to the affordances and conventions of different media, and as they learn to create using a range of different media tools.

• Students in literature classes are asked to take a familiar fairy tale, myth, or legend and identify how this story has been retold across different media, different historical periods, and different national contexts. Students search for recurring elements as well as signs of the changes that occur as the story is retold in a new context.

• French-language students in New York recreate characters from various French literary works in the best-selling video game *The Sims 2*. Students then tell new stories by playing out the interactions between different characters inside the game world. Characters are projected onto a screen in front of the class for students to do live performances with their characters.[105]

• An exercise developed by MIT's New Media Literacies project asks students to tell the same story across a range of different media. For example, they script dialogue using instant messenger; they storyboard using PowerPoint software and images appropriated from the Internet; later they might reenact their story and record it using a camera or video camera; they might illustrate it by drawing pictures. As they do so, they are encouraged to think about what each new tool contributes to their overall experience of the story as well as what needs to remain the same for viewers to recognize the same characters and situations across these various media.[106]

Networking: The Ability to Search for, Synthesize, and Disseminate Information

In a world in which knowledge production is collective and communication occurs across an array of different media, the capacity to network emerges as a core social skill and cultural competency. A resourceful student is no longer one who personally possesses a wide palette of resources and information from which to choose, but rather one who is able to successfully navigate an already abundant and continually changing world of information. Increasingly, students achieve this by tapping into a myriad of socially based search systems, including the following popular sites:

- Google.com: At the core of the now ubiquitous Google search engine, an algorithm analyzes the links between Web sites to measure which sites different Web site creators consider valuable or relevant to particular topics.
- Amazon.com: Suggests books a customer may like on the basis of patterns gleaned from analyzing similar customers.
- Movielens.org: Predicts if a particular user will like a given movie based on preferences from similar users.
- eBay.com: Creates a complex reputation system among users to establish trust for a given seller.
- Epinions.com: Establishes reliability of a given product on the basis of previous consumer experiences.
- Last.fm: Generates personalized radio stations on the basis of correlations between similar listeners' music preferences.
- Del.icio.us: Suggests relevant Web sites for a given term on the basis of other users' bookmarking habits.

- Answers.google.com: Offers a mass collective-intelligence marketplace in which users can offer money to anyone worldwide
who may have answers to their questions.
- Citeulike.org: This academic citation manager helps users
locate relevant articles on the basis of other users' citation man-
agement and allows users to flag important information about
given articles, such as inaccuracies.
- Getoutfoxed.com: Allows trusted friends and users to provide
annotations and metadiscussion about a given Web site that a
user might be browsing, such as warnings about inaccurate
content.
- RSS: Intelligently aggregates and consolidates content produced by friends and trusted sources to help efficiently share
resources across networks.

Business guru Tim O'Reilly has coined the term *Web 2.0* to
refer to how the value of these new networks depends not on
the hardware or the content, but on how they tap the participation of large-scale social communities that become invested in
collecting and annotating data for other users. Some of these
platforms require the active participation of consumers, relying
on a social ethos based on knowledge sharing. Others depend
on automated analysis of collective behavior. In both cases,
though, the value of the information depends on understanding how the knowledge is generated and analyzing the social
and psychological factors that shape collective behavior.

In such a world, students no longer can rely on expert gatekeepers to tell them what is worth knowing. Instead, they must
become more reflective about how individuals know what they
know and how they assess the motives and knowledge of differ-

ent communities. Students must be able to identify which group is most aware of relevant resources and choose a search system matched to the appropriate criteria: people with similar tastes; similar viewpoints; divergent viewpoints; similar goals; general popularity; trusted, unbiased, third-party assessment; and so forth. If transmedia navigation involves learning to understand the relations between different media systems, networking involves the ability to navigate across different social communities.

Schools are beginning to teach youths how to search out valuable resources through such activities as Webquests. The past ten years have seen an explosion in the popularity of these activities, designed by teachers, "in which some or all of the information that learners interact with comes from resources on the Internet."[107] In a typical Webquest, students arc given a scenario that requires them to extract information or images from a series of Web sites and then compile their findings into a final report. For example, students might be told they are part of a team of experts brought in to determine the most appropriate method for disposing of a canister of nuclear waste. They are provided a series of Web sites relevant to waste disposal and asked to present a final proposal to the teacher. For many educators, Webquests provide a practical means for using new

---

If transmedia navigation involves learning to understand the relations between different media systems, networking involves the ability to navigate across different social communities.

media to broaden students' exposure to different perspectives and provide fresh curricular materials. Rather than requiring textbook authors to develop "neutral" accounts of facts, teachers develop and share Webquests by simply referencing existing Web content. This both exposes students to a variety of opinions and trains them to synthesize their own perspectives. Yet critics argue that most existing Webquests fall short of fully exploring the potential of social networks—both in terms of teaching students how to exploit networking to track down information and in terms of using networks to distribute the byproducts of their research.

Networking is only partially about identifying potential resources; it also involves a process of synthesis, during which multiple resources are combined to produce new knowledge. In discussing "The Wisdom of Crowds," James Suroweicki describes the conditions needed to receive the maximum benefit from collective intelligence:

There are four key qualities that make a crowd smart. It needs to be diverse, so that people are bringing different pieces of information to the table. It needs to be decentralized, so that no one at the top is dictating the crowd's answer. It needs a way of summarizing people's opinions into one collective verdict. And the people in the crowd need to be independent, so that they pay attention mostly to their own information, and not worrying about what everyone around them thinks.[108]

Because new research processes depend on young people's resourcefulness as networkers, students must understand how to sample and distill multiple, independent perspectives. Guinee and Eagleton have been researching how students take notes in the digital environment, discovering, to their dismay, that young people tend to copy large blocks of text rather than para-phrasing it for future reference. In the process, they often lose track of the distinction between their own words and material borrowed from other sources.[109] They also skip over the need to assess any contradictions that might exist in the information they have copied. In short, they show only a minimal ability to create a meaningful synthesis from the resources they have gathered.

Networking also implies the youths' ability to effectively tap social networks to disperse their own ideas and media products. Many youths are creating independent media productions, but only some learn how to be heard by large audiences. Increasingly, young artists are tapping networks of fans or gamers with the goal of reaching a broader readership for their work.[110] They create within existing cultural communities not because they were inspired by a particular media property but because they want to reach that property's audience of loyal consumers. Young people are learning to link their Web sites together into Web rings in part to increase the visibility of any given site and also to increase the profile of the group. Teachers are finding that students are often more motivated if they can share what they create with a larger community. As students make their work accessible to a larger public, they face public consequences for what they write, and thus they face the kind of ethical dilemmas we identified earlier in this document.

At the present time, social networking software is under fire from adult authorities, and federal law makes it increasingly dif-ficult to access and deploy these tools in the classroom. Yet we would argue that schools have a different obligation—to help

> Schools have a different obligation—to help all children learn
> to use such tools effectively and to understand the value of
> networking as a means of acquiring knowledge and distributing
> information.

all children learn to use such tools effectively and to understand
the value of networking as a means of acquiring knowledge and
distributing information. Learning in a networked society
involves understanding how networks work and how to deploy
them to achieve particular ends. It involves understanding the
social and cultural contexts within which different information
emerges, when to trust and when not to trust others to filter
and prioritize relevant data, and how to use networks to get
individual work out into the world and in front of a relevant
and, with hope, appreciative public.

What Might Be Done

Educators take advantage of social networking when they link
learners with others who might share their interests or when
educators encourage students to publish their works to a larger
public.

▪ Noel Jenkins, a British junior-high teacher, created a geography unit in which students play the roles of city planners and
determine the most appropriate location for a new hospital in
San Francisco. First, students familiarize themselves with the
city layout by exploring satellite imagery of the city, navigating
through three-dimensional maps and watching webcam streams

from different parts of the city. Next, students learn how to layer the data that is most relevant to their decision atop their city maps. Finally, students must select a final location for the hospital and illustrate their maps with annotations justifying their decision.[111]

• Students use online storefront services such as CafePress and Zazzle to share their artistic creations and personal hobbies with the general public. In many cases, young entrepreneurs are able to make up to $18,000 per year doing so.[112]

• Educational technology enthusiast Will Richardson used the community news application CrispyNews to create www .edbloggernews.crispynews.com, an online nexus for teachers to share educational resources with one another. Each participant helps to rank the different curricular suggestions using collaborative filtering technologies.

• Students at Grandview Elementary School publish an online newspaper and podcast their works.[113]

• Outraged by a U.S. House of Representatives bill that would make illegal immigration a felony, more than 15,000 high school students in Los Angeles staged a protest coordinated primarily through MySpace.

Negotiation: The Ability to Travel across Diverse Communities, Discerning and Respecting Multiple Perspectives, and Grasping and Following Alternative Norms

The fluid communication within the new media environment brings together groups that otherwise might have segregated lives. Culture flows easily from one community to another. People online encounter conflicting values and assumptions and come to grips with competing claims about the meanings of shared artifacts and experiences. Everything about this process ensures that we will be provoked by cultural difference. Little about this process ensures that we will develop an understanding of the contexts within which these different cultural communities operate. When white suburban youths consume hip-hop or Western youths consume Japanese manga, new kinds of cultural understanding can emerge. Yet, just as often, the new experiences are read through existing prejudices and assumptions. Culture travels easily, but the individuals who initially produced and consumed such culture are not always welcome everywhere it circulates.

Cybercommunities often bring together groups that would have no direct contact in the physical world, resulting in heated conflicts about values or norms. Increasingly, critics are focusing on attempts to segregate membership or participation within online social groups. The massively multiplayer game *World of Warcraft* has faced controversies about whether the formation of groups for gay, lesbian, and bisexual players increased or decreased the likelihood of sexual harassment or whether the formation of groups based on English competency

---

The fluid communication within the new media environment brings together groups that otherwise might have segregated lives. Everything about this process ensures that we will be provoked by cultural difference. Little about this process ensures that we will develop an understanding of the contexts within which these different cultural communities operate.

reflected the importance of communication skills in games or constituted a form of discrimination motivated by stereotypes about the ethics and actions of Asian players. The social networking software that has become so central to youth culture can function as a vehicle for expressing and strengthening a sense of affiliation, but it also can be deployed as a weapon of exclusion and, as a consequence, as a tool for enforcing conformity to peer expectations.

In such a world, it becomes increasingly critical to help students acquire skills in understanding multiple perspectives, respecting and even embracing diversity of views, understanding a variety of social norms, and negotiating among conflicting opinions. Traditionally, media literacy has addressed these concerns by teaching children to read through media-constructed stereotypes about race, class, sex, ethnicity, religion, and other forms of cultural differences. Such work remains valuable in that it helps students to understand the preconceptions that may shape their interactions, but it takes on added importance as young people themselves create media content that may perpetuate stereotypes or contribute to misunderstandings. If, as writers such as Suroweicki and Lévy suggest, the wisdom of the crowd depends on the opportunity for diverse and independent insights and other inputs,[114] then these new knowledge cultures require participants to master new social skills that allow them to listen and respond to a range of different perspectives. We are defining this skill of negotiation in two ways: first, as the ability to negotiate between dissenting perspectives, and second, as the ability to negotiate through diverse communities.

The most meaningful interventions will start from a commitment to the process of deliberation and negotiation across dif-

It becomes increasingly critical to help students acquire skills in understanding multiple perspectives, respecting and even embracing diversity of views, understanding a variety of social norms, and negotiating between conflicting opinions.

ferences. They depend on the development of skills in active listening and of ethical principles designed to ensure mutual respect. Participants agree to some rules of conduct that allow them to talk through similarities and differences in perspective in ways that may allow for compromise, or at least an agreement to disagree. In either case, such an approach seems essential if we are to learn to share knowledge and collaborate within an increasingly multicultural society. Such an approach does not ignore differences—diversity of perspective is essential if the collective intelligence process is to work well—rather, it helps us to appreciate and value differences in background, experience, and resources as contributing to a richer pool of knowledge.

What Might Be Done
Educators can foster negotiation skills when they bring together groups from diverse backgrounds and provide them with resources and processes that ensure careful listening and deeper communication.

• Researchers at Stanford University's Center for Deliberative Democracy have been experimenting with new forms of civic engagement that depend on bringing people together from multiple backgrounds, exposing them to a broad array of perspectives, encouraging them to closely examine underlying claims and the evidence to support them, and creating a context in which they can learn from one another. Their initial reports suggest that this process generates powerful new perspectives on complex public-policy issues, which gain the support of all parties involved. For some participants, the process strengthens their commitment to core beliefs and values. For others, it creates a context in which they are more open to alternative points of view and are able to find middle-ground positions. The project's focus on the process of deliberation—and not simply on the outcome—represents a useful model to incorporate into the classroom. Rather than having traditional pro/con debates that depend on a fixed and adversarial relationship between participants, schools should focus more attention on group deliberation and decision-making processes and on mechanisms that ensure that all parties listen to and learn from one another's arguments.[115]

• *The Cultura Project*, developed by Gilberte Furstenberg, links students in classrooms in North America and France. In the first phases, they are asked to complete a series of sentences ("A good parent is someone who . . . "), address a series of questions ("What do you do if you see a mother strike a child in the grocery store?"), and define a range of core terms and concepts ("individualism"). The French students write in French and the American students in English, allowing both classes to practice their language skills and understand the links between linguistic and cultural practices. Students then are asked to compare the different ways that people living in different parts of the world responded to these questions, seeking insights into differences in values and lifestyles. For example, individualism in

France is seen as a vice, equated with selfishness, whereas for Americans individualism is seen as a virtue, closely linked with freedom. These interpretations unfold in online forums, where students from both countries can respond to and critique attempts to characterize their attitudes. As the process continues, students are encouraged to upload their own media texts, which capture important aspects of their everyday lives, as artifacts they believe speak to the larger cultural questions at the center of their discussions. In this way, they learn to see themselves and one another more clearly, and they come away with a greater appreciation of cultural difference.[116]

• Rev. Denis Haak of the Ransom Fellowship has developed a series of probing questions and exercises intended to help Christians work through their responses to popular culture. Rejecting a culture-war rhetoric based on sharp divisions, these exercises are intended to help Christians to identify and preserve their own values even as they come to understand "what non-believers believe." The Discernment movement sees discussing popular culture as a means of making sense of competing and contradictory value systems that interact in contemporary society. For this process to work, the program encourages participants to learn how to "disagree agreeably," to stake out competing positions without personalizing the conflict.[117]

• Schools historically have used the adversarial process of formal debate to encourage students to conduct research, construct arguments, and mobilize evidence. Yet there is a danger that this process forces students to adopt fixed and opposing positions on complex problems. A better approach might be to adopt a deliberative process in the classroom that encourages

> Literacy skills for the twenty-first century are skills that enable participation in the new communities emerging within a networked society. They enable students to exploit new simulation tools, information appliances, and social networks; they facilitate the exchange of information among diverse communities and the ability to move easily across different media platforms and social networks.

collaboration and discussion across different positions and thus creates a context for opposing parties to learn from one another and reformulate their positions accordingly.

- Sites such as Wikipedia and Wikinews include a tab labeled *discussion* above each article or news entry. Here readers can view or participate in an online discussion with people of different viewpoints to arrive at a neutral-point-of-view framing of the issue to be displayed on the main page.

We began this discussion by suggesting that literacy in the twenty-first century be understood as a social rather than individual skill and that what students must acquire should be understood as skills and cultural competencies. Each of the skills we have identified above represents modes of thought, ways of processing information, and ways of interacting with others to produce and circulate knowledge. These are skills that enable participation in the new communities emerging within a networked society. They enable students to exploit new simulation tools, information appliances, and social networks; they facilitate the exchange of information among diverse communities and the ability to move easily across different media plat-forms and social networks. Many of these skills schools have been teaching all along, although the emergence of digital media creates new pressure to prepare students for their future roles as citizens and workers. Others are skills that emerge from the affordances of these new communications technologies and the social communities and cultural practices that have grown up around them.

Who Should Respond? A Systemic Approach to Media Education

We have identified three core problems that should concern everyone who cares about the development and well-being of America's youth:

- How do we ensure that every child has access to the skills and experiences needed to become a full participant in the social, cultural, economic, and political future of our society?
- How do we ensure that every child has the ability to articulate his or her understanding of the way that media shapes perceptions of the world?
- How do we ensure that every child has been socialized into the emerging ethical standards that will shape their practices as media makers and as participants within online communities?

We have also identified a set of core social skills and cultural competencies that young people should acquire if they are to be full, active, creative, and ethical participants in this emerging participatory culture:

Play The capacity to experiment with the surroundings as a form of problem solving.

Performance   The ability to adopt alternative identities for the purpose of improvisation and discovery.

Simulation   The ability to interpret and construct dynamic models of real-world processes.

Appropriation   The ability to meaningfully sample and remix media content.

Multitasking   The ability to scan the environment and shift focus onto salient details.

Distributed cognition   The ability to interact meaningfully with tools that expand mental capacities.

Collective intelligence   The ability to pool knowledge and compare notes with others toward a common goal.

Judgment   The ability to evaluate the reliability and credibility of different information sources.

Transmedia navigation   The ability to follow the flow of stories and information across multiple modalities.

Networking   The ability to search for, synthesize, and disseminate information.

Negotiation   The ability to travel across diverse communities, discerning and respecting multiple perspectives, and grasping and following alternative norms.

Some children acquire some of these skills through their participation in the informal learning communities that surround popular culture. Some teachers incorporate some of these skills into their classroom instruction. And some after-school programs infuse some of these skills into their activities. Yet, as the above qualifications suggest, the integration of these important social skills and cultural competencies remains haphazard at best. Media education is taking place for some youths across a

> The integration of important social skills and cultural competencies remains haphazard at best. Media education is taking place for some youth across a variety of contexts, but it is not a central part of the educational experience of all students.

variety of contexts, but it is not a central part of the educational experience of all students. Our goal for this report is to encourage greater reflection and public discussion on how we might incorporate these core principles systematically across curricula and across the divide between in-school and out-of-school activities. Such a systemic approach is needed if we are to close the participation gap, confront the transparency problem, and help young people work through the ethical dilemmas they face in their everyday lives. Such a systemic approach is needed if children are to acquire the core social skills and cultural competencies needed in a modern era.

Schools

In the above descriptions of core social skills and cultural competencies, we have spotlighted a range of existing classroom practices that help children become fuller participants in the new media landscape: the use of educational simulations, alternative and augmented reality games, Webquests, production activities, blogs and wikis, and deliberation exercises. Such exercises involve actively applying new techniques of knowledge production and community participation to the existing range of academic subjects in the established school curriculum. We have seen how history classes are making use of educational games, how science classes are teaching youths to evaluate and construct simulations, how literature classes are embracing role-play and appropriation, how math classes might explore the value of distributed cognition, and how foreign language classes are bridging cultural differences via networking. As these examples suggest, many individual schools and educators are experimenting with new media technologies and the processes of collaboration, networking, appropriation, participation, and expression that they enable. Real-world inquiries require students to search out information, interview experts, connect with other students around the world, generate and share multimedia, assess digital documents, write for authentic audiences, and otherwise exploit the resources of the new participatory culture.

We see this report as supporting these individual educators by encouraging a more systemic consideration of the place these skills should assume in pedagogical practice. We believe that these core social skills and cultural competencies have implications across the school curriculum, with each teacher assuming responsibility for helping students develop the skills necessary for participation within their discipline. Clearly, more discipline-specific research is needed to fully understand the value and relevance of these skills to different aspects of the school curriculum. Skills that are already part of the professional practices of scientists, historians, artists, and policymakers can also help inform how we introduce students to these disciplines.

Much of the resistance to media literacy training springs from the sense that the school day is bursting at its seams and we

Much of the resistance to media literacy training springs from the sense that the school day is bursting at its seams and we cannot cram in any new tasks without the instructional system breaking down altogether. For that reason, we do not want to see media literacy treated as an add-on subject. Rather, its introduction should be a paradigm shift that, like multiculturalism or globalization, reshapes how we teach every existing subject.

cannot cram in any new tasks without the instructional system breaking down altogether. For that reason, we do not want to see media literacy treated as an add-on subject. Rather, its introduction should be a paradigm shift that, like multiculturalism or globalization, reshapes how we teach every existing subject. Media change is affecting every aspect of our contemporary experience and, as a consequence, every school discipline needs to take responsibility for helping students to master the skills and knowledge they need to function in a hypermediated environment.

After School

After-school programs may encourage students to examine more directly their relationships to popular media and participatory culture. After-school programs may introduce core technical skills that students need to advance as media makers. In these more informal learning contexts, students may explore rich examples of existing media practice and develop a vocabulary for critically assessing work in these emerging fields. Students also may have more time to produce their own media and to reflect on their own production activities. The approach proposed here takes the best of several contemporary approaches to media education, fusing the critical skills and inquiry associated with media literacy with the production skills associated with Computer Clubhouses (discussed below), and adding to both a greater awareness of the politics and practice of participatory culture.

The media literacy movement emerged in response to the rise of mass media. Here, for example, is a classic definition of *media literacy* created by the Ontario Association for Media Literacy in 1989:

Media literacy is concerned with developing an informed and critical understanding of the nature of the mass media, the techniques used by them, and the impact of those techniques. It is education that aims to increase students' understanding and enjoyment of how the media work, how they produce meaning, how they are organized, and how they construct reality. Media literacy also aims to provide students with the ability to create media products.[118]

Although some media literacy educators have instituted groundbreaking work on digital media, the bulk of presentations at national conferences are still focused on more traditional media—print, broadcast, cinema, popular music, advertising— which are assumed to exert the greatest influence on young people's lives.

Media literacy educators are not wrong to be concerned by the concentrated power of the media industry, but they also must realize that this is only part of a more complex picture. We live in a world in which media power is more concentrated

Existing media literacy materials give us a rich vocabulary for thinking about issues of representation, helping students to think critically about how the media frames perceptions of the world and reshapes experience according to its own codes and conventions. Yet these concepts need to be rethought for an era of participatory culture.

than ever before, yet the ability of everyday people to produce and distribute media has never been freer. Existing media literacy materials give us a rich vocabulary for thinking about issues of representation, helping students to think critically about how the media frames perceptions of the world and reshapes experience according to its own codes and conventions. Yet these concepts need to be rethought for an era of participatory culture.

Consider, for example, the framework for media literacy proposed by Share, Jolls, and Thoman:

- Who created the message?
- What creative techniques are used to attract my attention?
- How may different people understand this message differently than me?
- What lifestyles, values, and points of view are represented in—or omitted from—this message?
- Why is this message being sent?[119]

There is much to praise in these questions: they understand media as operating within a social and cultural context, they recognize that what we take from a message is different from what the author intended, they focus on interpretation and context as well as motivation, and they are not tied up with a language of victimization.

Yet note that each question operates on the assumption that the message was created elsewhere and that we are simply its recipients (critical, appropriating, or otherwise). We would add new complexity and depth to each of these questions if we rephrased them to emphasize individuals' own active participation in selecting, creating, remaking, critiquing, and circulating media content. One of the biggest contributions of the media literacy movement has been this focus on inquiry, identifying core questions that can be asked of a broad range of different media forms and experiences. This inquiry process seems key to overcoming the transparency problems identified above.

By contrast, education for the digital revolution stressed tools above all else. The challenge was to wire the classroom and prepare youths for the demands of the new technologies. Computer Clubhouses sprang up around the country to provide learning environments where youths could experiment with new media techniques and technologies. The goal was to allow students to set and complete their own tasks with the focus almost entirely on the production process. Little effort was made to give youths a context for thinking about these changes or to reflect on the new responsibilities and challenges they faced as participants in the digital culture. We embrace the constructivist principles that have shaped the Computer Clubhouse movement: youths do their best work when engaged in activities that are personally meaningful to them. Yet we also see a value in teaching youths how to evaluate their own work and

We embrace the constructivist principles that have shaped the Computer Clubhouse movement: youth do their best work when engaged in activities that are personally meaningful to them. Yet we also see a value in teaching youth how to evaluate their own work and appraise their own actions, and we see the necessity of helping them to situate the media they produce within its larger social, cultural, and legal context.

appraise their own actions, and we see the necessity of helping them to situate the media they produce within its larger social, cultural, and legal context.

We have developed an integrated approach to media pedagogy founded on exercises that introduce youths to core technical skills and cultural competencies, exemplars that teach youths to critically analyze existing media texts, expressions that encourage youths to create new media content, and ethics that encourage youths to critically reflect on the consequences of their own choices as media makers.

School-based and after-school programs serve distinct but complementary functions. We make a mistake when we use after-school programs simply to play catch-up on school-based standards or to reinforce what schools are already teaching. After-school programs should be a site of experimentation and innovation, a place where educators catch up with the changing culture and teach new subjects that expand children's understanding of the world. After-school programs focused on media education should function in a variety of contexts. Museums, public libraries, churches, and social organizations (such as the YWCA or the Boy Scouts) can play important roles, each drawing on its core strengths to expand beyond what can be done during the official school day.

Parents

We also see an active role for parents to play in shaping children's earliest relationships to media and reinforcing their emerging skills and competencies. The new media technologies give parents greater control over the flow of media into their lives than ever before, yet parents often describe themselves as overwhelmed by the role that media plays in their children's everyday activities. As *UK Children Go On-line* concluded, "Opportunities and risks go hand in hand. . . . The more children experience one, the more they also experience the other."[120] Rather than constraining choices to protect youths from risks, the report advocates doing a better job of helping youths master the skills they need to exploit opportunities and avoid pitfalls.

Parents lack basic information that would help them deal with both the expanding media options and the breakdown of traditional gatekeeping functions. Most existing research focuses on how to minimize the risks of exposure to media, yet we have stressed the educational benefits of involvement in participatory culture. The first five or six years of a child's life are formative for literacy and social skills, and parents can play an important part in helping children acquire the most basic versions of the skills we have described here. Throughout children's lives, parents play important roles in helping them make mean-

Adults often are led by fears and anxieties about new forms of media that were not a part of their own childhood and which they do not fully understand. There are few, if any, books that offer parents advice on how to make these choices or provide information about the media landscape.

ingful choices in their use of media and in helping them anticipate the consequences of the choices they make. Adults often are led by fears and anxieties about new forms of media that were not a part of their own childhood and which they do not fully understand. There are few, if any, books that offer parents advice on how to make these choices or provide information about the media landscape. Few education programs help parents to acquire skills and self-confidence to help their children master the new media literacies, and few sites provide up-to-date and ongoing discussions of some of the issues surrounding the place of media in children's lives.

The Challenge Ahead: Ensuring that All Benefit from the Expanding Media Landscape

Writing in the *Chronicle of Higher Education*, Bill Ivey, the former chairman of the National Endowment for the Arts, and Steven J. Tepper, a professor of sociology at Vanderbilt University, described what they see as the long-term consequences of this participation gap:

Increasingly, those who have the education, skills, financial resources, and time required to navigate the sea of cultural choice will gain access to new cultural opportunities. . . . They will be the pro-ams [professional amateurs] who network with other serious amateurs and find audiences for their work. They will discover new forms of cultural expression that engage their passions and help them forge their own identities, and will be the curators of their own expressive lives and the mavens who enrich the lives of others. . . . At the same time, those citizens who have fewer resources—less time, less money, and less knowledge about how to navigate the cultural system—will increasingly rely on the cultural fare offered to them by consolidated media and entertainment conglomerates. . . . Finding it increasingly difficult to take advantage of the pro-am revolution, such citizens will be trapped on the wrong side of the cultural divide. So technology and economic change are conspiring to create a new cultural elite—and a new cultural underclass. It is not yet clear what such a cultural divide portends: what its consequences will be for democracy, civility, community, and quality of life. But the emerging picture is deeply troubling. Can America prosper if its citizens experience such different and unequal cultural lives?[121]

Ivey and Tepper bring us back to the core concerns that have framed this essay: how can we "ensure that all students benefit from learning in ways that allow them to participate fully in public, community, [creative,] and economic life?" How do we guarantee that the rich opportunities afforded by the expanding media landscape are available to all? What can we do in schools, after-school programs, and the home to give our youngest children a head start and allow our more mature youths the chance to develop and grow as effective participants and ethical communicators? This is the challenge that faces education at all levels at the dawn of a new era of participatory culture.

Notes

1. New London Group, "A Pedagogy of Multiliteracies: Designing Social Futures," in *Multiliteracies: Literacy Learning and the Design of Social Futures*, ed. Bill Cope and Mary Kalantzis (London: Routledge, 2000), 9–38.

2. Henry Jenkins, "Playing Politics in Alphaville," *Technology Review* (May 7, 2004), http://www.technologyreview.com/Hardware/wtr_136 06,294,p1.html.

3. Henry Jenkins, *Convergence Culture: Where Old and New Media Collide* (New York: New York University Press, 2006).

4. Josh McHugh, "The Firefox Explosion," *Wired Magazine* 13.02 (February 2005), http://www.wired.com/wired/archive/13.02/firefox_pr.html.

5. Vanessa Bertozzi and Henry Jenkins, *Young Artists* (New York: Routledge, forthcoming).

6. Lenhart and Madden, *Teen Content.*

7. Sonia Livingstone, *The Changing Nature and Uses of Media Literacy* (working paper, London School of Economics, 2003), 15–16, http://www.lse.ac.uk/collections/media@lse/mediaWorkingPapers/ewpNumber4.htm (accessed September 2006).

8. Lisa Gitelman, *Scripts, Grooves, and Writing Machines: Representing Technology in the Edison Era* (Stanford, CA: Stanford University Press, 1999).

9. Jenkins, *Convergence Culture.*

10. James Paul Gee, *Situated Language and Learning: A Critique of Traditional Schooling* (New York: Routledge, 2004).

11. Rebecca W. Black, "Access and Affiliation: The Literacy and Composition Practices of English Language Learners in an Online Fanfiction Community," *Journal of Adolescent & Adult Literacy* 49, no. 2 (2005): 118–128; Rebecca W. Black, "Online Fanfiction: What Technology and Popular Culture Can Teach Us about Writing and Literacy Instruction" *New Horizons for Learning Online Journal* 11, no. 2 (Spring 2005).

12. Andrew Blau, "The Future of Independent Media," *Deeper News* 10, no. 1 (2005): 3, http://www.gbn.com/ArticleDisplayServlet.srv?aid=34045.

13. Lenhart and Madden, *Teen Content.*

14. David Buckingham, *The Making of Citizens: Young People, News and Politics* (London: Routledge, 2000), 218–219.

15. John C. Beck and Mitchell Wade, *Got Game? How the Gamer Generation Is Reshaping Business Forever* (Cambridge, MA: Harvard Business School Press, 2004).

16. Kaiser Family Foundation, *Generation M: Media in the Lives of 8–18 Year Olds* (March 9, 2005), http://www.kff.org/entmedia/entmedia030905pkg.cfm (accessed September 2006); Kaiser Family Foundation, *The Effects of Electronic Media on Children Ages Zero to Six: A History of Research* (January 2005), http://www.kff.org/entmedia/upload/The-Effects-of-Electronic-Media-on-Children-Ages-Zero-to-Six-A-History-of-Research-Issue-Brief.pdf.

17. PBS *Nightly News Hour*, transcripts, November 22, 2005, http://www.pbs.org/newshour/bb/cyberspace/july-dec05/philadelphia_11-22.html.

18. Ibid.

19. Sonia Livingstone and Magdalena Bober, *UK Children Go Online* (London: Economic and Social Research Council, 2005), 12, http://personal.lse.ac.uk/bober/UKCGOfinalReport.pdf.

20. Current legislation to block access to social networking software in schools and public libraries will further widen the participation gap.

21. Ellen Wartella, Barbara O'Keefe, and Ronda Scantlin, *Children and Interactive Media: A Compendium of Current Research and Directions for the Future* (New York: Markle Foundation, 2000), 8, http://www.markle.org/downloadable_assets/cimcompendium.pdf.

22. Peter Lyman and others, "Literature Review: Digital-Mediated Experiences and Youth's Informal Learning" (report, Exploratorium, San Francisco, 2005), http://www.exploratorium.edu/researc/digitalkids/Lyman_DigitalKids.pdf.

23. Manuel Castells, *The Internet Galaxy: Reflections of the Internet, Business, and Society* (Oxford: Oxford University Press, 2002), quoted in Livingstone, *Changing Nature.*

24. Gee, *Situated Language*, 105.

25. Sherry Turkle, *Life on the Screen: Identity in the Age of the Internet* (New York: Simon & Schuster, 1995), 70.

26. Ted Friedman, "Making Sense of Software: Computer Games and Interactive Textuality," in *Cybersociety*, ed. Steven G. Jones (Thousand Oaks, CA: Sage, 1995).

27. Karen L. Schrier, "Revolutionizing History Education: Using Augmented Reality Games to Teach History" (master's thesis, Comparative Media Studies Program, Massachusetts Institute of Technology, 2005).

28. Kurt Squire, "Replaying History: Learning World History through Playing *Civilization III*" (PhD diss., Instructional Systems and Technology Department, Indiana University, 2004).

29. Renee Hobbs, "Deciding What to Believe in an Age of Information Abundance: Exploring Non-Fiction Television in Education," *Sacred Heart Review* 42 (1999): 4–26, http://www.reneehobbs.org/renee's%20web%20site/Publications/Sacred%20Heart.htm (accessed September 2006).

30. Howard Gardner, interview with author, 2006; see also Peter Levine, "The Audience Problem," in *Digital Media and Youth Civic Engagement*, ed. W. Lance Bennett (Cambridge, MA: MIT Press, forthcoming).

31. Ellen Seiter, *The Internet Playground: Children's Access, Entertainment, and Mis-Education* (London: Peter Lang, 2005): 38.

32. Ibid., 100.

33. Wendy Fischman, Becca Solomon, Deborah Greenspan, and Howard Gardner, *Making Good: How Young People Cope with Moral Dilemmas at Work* (Cambridge, MA: Harvard University Press, 2004).

34. Julian Dibbell, "A Rape in Cyberspace, or How an Evil Clown, a Haitian Trickster Spirit, Two Wizards, and a Cast of Dozens Turned a Database into a Society," *The Village Voice* (December 21, 1993): 36–42; Henry Jenkins, "Playing Politics"; always_black, "Bow Nigger," (2004), http://www.alwaysblack.com/blackbox/bownigger.html (accessed March 17, 2006).

35. Bertram C. Bruce, "Diversity and Critical Social Engagement: How Changing Technologies Enable New Modes of Literacy in Changing Circumstances," in *Adolescents and Literacies in a Digital World*, ed. Donna E. Alvermann (New York: Peter Lang, 2002).

36. New Media Consortium, *A Global Imperative: The Report of the 21st Century Literacy Summit* (New Media Consortium, 2005), 8.

37. Black, "Access and Affiliation"; Jenkins, *Convergence Culture*.

38. New Media Consortium, *Global Imperative*.

39. New London Group, "A Pedagogy of Multiliteracies," 9.

40. Mary Louise Pratt, "Arts of the Contact Zone," *Profession* 91 (1991): 33–40.

41. Ibid., 34.

42. Henry Jenkins, "Fun vs. Engagement: The Case of the Zoobinis," *Confessions of an Aca-Fan* (June 23, 2006), http://www.henryjenkins .org/2006/06/fun_vs_engagement_the_case_of.html#more.

43. James P. Gee, *What Video Games Can Teach Us about Literacy and Learning* (New York: Palgrave-McMillan, 2003).

44. Henry Jenkins, "Buy These Problems Because They're Fun to Solve! A Conversation with Will Wright," *Telemedium: The Journal of Media Literacy* 52 nos. 1 and 2 (2005): 21.

45. Squire, "Replaying History."

46. Seth Kahan, "John Seely Brown" (February 10, 2003), http://www .sethkahan.com/Resources_03BrownJohnSeely.html.

47. Eric Klopfer, personal correspondence with author.

48. Ian Bogost, "Procedural Literacy: Problem Solving with Programming, Systems and Play," *Telemedium: The Journal of Media Literacy* 52 nos. 1 and 2 (2005): 36.

49. Andy Clark, *Natural-Born Cyborgs: Minds, Technologies, and the Future of Human Intelligence* (Oxford: Oxford University Press, 2003), 160.

50. Ibid.

51. James Paul Gee, *Video Games*, 55.

52. Russell Francis, "Towards a Theory of a Games Based Pedagogy" (paper presented at the Innovating e-Learning 2006: Transforming Learning Experiences, JISC Online Conference, United Kingdom, 2006).

53. Ibid.

54. Shelby A. Wolf and Shirley B. Heath, *The Braid of Literature: Children's Worlds of Reading* (Cambridge, MA: Harvard University Press, 1992).

55. Anne Haas Dyson, *Writing Superheroes: Contemporary Childhood, Popular Culture, and Classroom Literacy* (New York: The Teachers College Press, 1997).

56. Susannah Stern, "Growing Up Online," *Telemedium: The Journal of Media Literacy* 52 nos. 1 and 2 (2005): 57.

57. Bertozzi and Jenkins, *Young Artists.*

58. David Williamson Shaffer, "Epistemic Frames for Epistemic Games," *Computers and Education* (2005), http://coweb.wcer.wisc.edu/cv/papers/Shaffer_cae_2005.pdf.

59. For more information, see http://web.archive.org/web/2006090 5192731/http://www.incsub.org/wpmu/bionicteacher/?p=142.

60. Lenhart and Madden, *Teen Content.*

61. Henry Jenkins, "The MIT Games Literacy Workshop," *Telemedium: The Journal of Media Literacy* 52 nos. 1 and 2 (2005): 37–40.

62. Jenkins, "Fun vs. Engagement."

63. Ibid.

64. Ibid.

65. Alan D. Baddeley, *Essentials of Human Memory* (East Sussex, UK: Psychology Press, 1999).

66. John Seely Brown, "Growing Up Digital: How the Web Changes Work, Education, and the Ways People Learn," *Change* (March/April 2000): 10–20, http://www.usdla.org/html/journal/FEB02_Issue/article01.html (accessed September 2006).

67. Henry Jenkins, "Video Game Virtue," *Technology Review Online* (August 1, 2003), http://www.technologyreview.com/articles/03/08/wo_jenkins080103.asp?p=1.

68. Gunther Kress, *Literacy in the New Media Age* (New York: Routledge, 2003).

69. Thom Hartmann, *Attention Deficit Disorder: A Different Perception* (New York: Gill & MacMillan, 1999).

70. Mark Glassman, "Maroon 5 Makes Room on the IPod for School-work," *New York Times*, December 9, 2004, http://www.nytimes.com/2004/12/09/technology/circuits/09ipod.html (accessed May 2009).

71. Andy Clark, *Being There: Putting Brain, Body, and World Together Again* (Cambridge, MA: MIT Press, 1997).

72. Clark, *Natural-Born Cyborgs*.

73. Roy Pea, "Practices of Distributed Intelligence and Designs for Education," in *Distributed Cognitions: Psychological and Educational Considerations*, ed. Gavriel Salomon (Cambridge: Cambridge University Press, 1993), 50.

74. David Williamson Shaffer and James J. Kaput, "Mathematics and Virtual Culture: An Evolutionary Perspective on Technology and Mathematics," *Educational Studies in Mathematics* 37 (1999): 97–119.

75. Clark, *Natural-Born Cyborgs*.

76. Gee, *Video Games*.

77. Eric Klopfer and Kurt Squire, "Environmental Detectives: The Development of an Augmented Reality Platform for Environmental Simulations," *Educational Technology Research and Development* 56 no. 2 (April 2008): 203–228.

78. Henry Jenkins, "Look, Listen, Walk," *Technology Review Online* (April 2, 2004), http://www.technologyreview.com/InfoTech/wtr_13560,294,p1.html.

79. Schrier, "Revolutionizing History Education."

80. David Hatfield and David Williamson Shaffer, "Press Play: Designing an Epistemic Game Engine for Journalism," in *Proceedings of the 7th International Conference on Learning Sciences* (New York: ACM Press, 2006), 236–242.

81. Philip Bell and William Winn, "Distributed Cognition, by Nature and by Design," in *Theoretical Foundations of Learning Environments*, ed. David H. Jonassen & Susan M. Land (Mahwah, NJ: Lawrence Erlbaum, 2000), 123–145.

82. David Williamson Shaffer and Katherine A. Clinton, "Tool-forthoughts: Reexamining Thinking in the Digital Age," *Mind, Culture, and Activity* 13 no. 4 (2006): 283–300, http://epistemicgames.org/cv/papers/toolforthoughts-sub1.pdf.

83. Robert P. Yagelski, "Computers, Literacy, and Being: Teaching with Technology for a Sustainable Future," http://english.ttu.edu/kairos/6.2/features/yagelski/crisis.htm.

84. Pierre Lévy, *Collective Intelligence: Man's Emerging World in Cyberspace* (New York: Perseus, 2000).

85. Jane McGonigal, "This Is Not a Game: Immersive Aesthetics and Collective Play," *Digital Arts & Culture 2003 Conference Proceedings* (2003), http://www.seanstewart.org/beast/mcgonigal/notagame/paper.pdf.

86. Jenkins, *Convergence Culture*.

87. McGonigal, "Not a Game," 7.

88. Cory Doctorow, "Digital Utopia and Its Flaws," *Neofiles* 1 no. 15 (2005) http://www.life-enhancement.com/NeoFiles/default.asp?id=23 (accessed September 2006).

89. Ryan Singel, "A Disaster Map 'Wiki' is Born," *Wired News* (2005), http://www.wired.com/news/technology/0,68743-0.html.

90. Sally Atwood, "Education Arcade: MIT Researchers Are Creating Academically Driven Computer Games That Rival Commercial Products and Make Learning Fun," *Technology Review* (June 12, 2004), http://www.matr.net/article-11198.html.

91. See http://seagrantdev.mit.edu/eelgrass/.

92. Jim Giles, "Internet Encyclopaedias Go Head to Head," *Nature* 438 (December 15, 2005): 900–901, http://www.nature.com/nature/journal/v438/n7070/full/438900a.html.

93. John Seigenthaler, "A False Wikipedia 'Biography.'" *USAToday*, November 29, 2005, Op-ed, http://www.usatoday.com/news/opinion/editorials/2005-11-29-wikipedia-edit_x.htm.

94. Dan Gillmor, *We the Media* . . . (New York: O'Reilly Media, 2004).

95. Jeff Share, Tessa Jolls, and Elizabeth Thoman, *Five Key Questions That Can Change the World* (San Francisco: Center for Media Literacy, 2005), 182.

96. David Buckingham, "The Media Literacy of Children and Young People: A Review of the Literature" (report, Centre for the Study of Children, Youth and Media, Institute of Education, University of London, 2005), 22, http://www.ofcom.org.uk/advice/media_literacy/medlitpub/medlitpubrss/ml_children.pdf (accessed September 2006).

97. Keri Facer, John Furlong, Ruth Furlong, and Rosamund Sutherland, *ScreenPlay: Children and Computing in the Home* (London: Routledge Falmer, 2003), quoted in Buckingham, "Media Literacy," 18.

98. Jenkins, *Convergence Culture*.

99. David Buckingham and Julian Sefton-Green, "Structure, Agency, and Pedagogy in Children's Media Culture," in *Pikachu's Global Adventure: The Rise and Fall of Pokémon,* ed. Joseph Tobin (Durham, NC: Duke University Press, 2004), 22.

100. Mizuko Ito, "Technologies of the Childhood Imagination: Yugioh, Media Mixes, and Everyday Cultural Production," in *Network/Netplay: Structures of Participation in Digital Culture,* ed. Joe Karaganis and Natalie Jeremijenko (Durham, NC: Duke University Press, 2005).

101. Kress, *Literacy in the New Media Age.*

102. Ibid.

103. New Media Consortium, *Global Imperative.*

104. James Daly, "Life on the Screen," *Edutopia* (2004), http://www .edutopia.org/magazine/ed1article.php?id=art_1160 (accessed September 2006).

105. See http://www.mylenecatel.com.

106. Jenkins, "Fun vs. Engagement."

107. Bernie Dodge, "Some Thoughts about Webquests," 1997, http:// webquest.sdsu.edu/about_webquests.html (accessed September 2006).

108. James Suroweicki, "The Wisdom of Crowds: Q & A with James Suroweicki," New York: Random House Web site, 2004, http://www .randomhouse.com/features/wisdomofcrowds/Q&A.html (accessed September 2006).

109. Kathleen Guinee and Maya B. Eagleton, "Spinning Straw into Gold: Transforming Information into Knowledge during Web-based Research," *English Journal* 95 no. 4 (2006).

110. Bertozzi and Jenkins, *Young Artists.*

111. See Noel Jenkins, "San Francisco: Visualizing a Safer City," *Juicy Geography* 2006, http://www.juicygeography.co.uk/googleearthsanfran .htm.

112. Bobbi Barbour, "DollFacePunk in the News," 2006, http://doll facepunk.com/Pres.htm.

113. See http://www.grandviewlibrary.org/Fold/GrandviewNews.aspx.

114. Suroweicki, "The Wisdom of Crowds"; Lévy, *Collective Intelligence*.

115. James Fishkin and Robert C. Lushkin, "Experimenting with a Democratic Ideal: Deliberative Polling and Public Opinion" (paper, Swiss Chair's Conference on Deliberation, the European University Institute, Florence, Italy, May 21–22, 2004), http://cdd.stanford.edu/research/papers/2004/democraticideal.pdf.

116. Gilberte Furstenberg, *The Cultura Project* (Washington, DC: National Capital Language Resource Center Culture Club, 2004), http://www.nclrc.org/profdev/cultureclub0407.html (accessed September 2006).

117. Jenkins, *Convergence Culture*.

118. Quoted in Barry Duncan, "Media Literacy: Essential Survival Skills for the New Millennium," *Orbit Magazine* 35 no. 2 (2005), http://www.oise.utoronto.ca/orbit/mediaed_sample.html.

119. Share, Jolls, and Thoman, *Five Key Questions*.

120. Livingstone and Bober, *UK Children*.

121. Bill Ivey and Steven J. Tepper, "Cultural Renaissance or Cultural Divide?" *Chronicle of Higher Education* 52 (May 19, 2006).